THE THIRD DOMAIN

THE UNTOLD STORY OF ARCHAEA AND THE FUTURE OF BIOTECHNOLOGY

THE THIRD DOMAIN

THE UNTOLD STORY OF ARCHAEA AND THE FUTURE OF BIOTECHNOLOGY

TIM FRIEND

Joseph Henry Press
Washington, D.C.

Joseph Henry Press 500 Fifth Street, NW Washington, DC 20001

The Joseph Henry Press, an imprint of the National Academies Press, was created with the goal of making books on science, technology, and health more widely available to professionals and the public. Joseph Henry was one of the founders of the National Academy of Sciences and a leader in early American science.

Library of Congress Cataloging-in-Publication Data

Friend, Tim.
The third domain : the untold story of archaea and the future of biotechnology / Tim Friend.
p. cm.
Includes bibliographical references and index.
ISBN 978-0-309-10237-7 (hardcover)
1. Archaeobacteria—History. 2. Archaeobacteria—Biotechnology.
I. Title.
QR82.A69F75 2007
579.3′21—dc22

2007006291

Cover design by Michele de la Menardiere.

Printed in the United States of America

To Robin, for a new path.

CONTENTS

1
LIFE IN THE OUTER LIMITS

And now for something completely different.
Monty Python

The microbial stench of the Harlem Meer is soaked into our wetsuits, even our skin. Long green strands of algae cling to our heads, shoulders, and arms as we rise from the water, giving the appearance and presumably the odor of sea monsters. Ever see *Creature from the Black Lagoon*? "Not since the beginning of time has the world beheld terror like this!" Yeah, well. Go scuba diving in Central Park.

I have just finished collecting from the Harlem Meer the last samples of microorganisms for this mostly science and partly adventure tale about the discovery of a strange form of microbial life called the archaea. The Harlem Meer is a small lake in the northeast corner of Central Park. It is by far the nastiest water in which I've had the pleasure to dive. Water leaking around the mouthpiece of my regulator has left a lingering, tangy taste, and perhaps a burning sensation in the back of my throat. The palate suggests hydrocarbons, metals, and something awfully organic. No doubt the samples of filthy sludge and turbid water collected into ziplock baggies and plastic bottles will contain many interesting microbes.

The archaea—pronounced "ar-kee-ah"—have been found thriving in every conceivable environment on Earth. Discovered only 30 years ago, their remarkable adaptability and versatile metabolisms make them perhaps the most formidable life on the planet. Initially, scientists found colonies of archaea in all of the extreme places where life is not supposed to exist, such as boiling pools of sulfuric acid around volcanoes, glacial ice, and even toxic waste dumps. Quite recently, scientists who specialize in these microbes were astounded to discover that archaea also are as abundant in our soils, oceans, and lakes as bacteria, and, since bacteria were believed to be solely responsible for helping plants grow and generating at least half of the oxygen we breathe, this surprising discovery of plentiful archaea should send the scientists who study climate change and global warming back to the drawing board to recalculate archaea's role in the fundamental cycle of life. The problem is, most of them don't know it yet.

No one is sure yet whether archaea can cause diseases, but they were recently discovered colonizing our gums in some type of collusion with the infectious bacteria that can rot our teeth.[1] As the fame of archaea spreads into the biomedical community, more associations with diseases are likely to come.

It seems incredible that such a common and presumably vital form of life, however strange in its habits, could remain invisible to science for so long. But, part of the reason has to do with the fact that "mainstream" scientists thought they had life on Earth all figured out. Life could not possibly exist where scientists did not believe it could exist. When one pioneering biophysicist named Carl Woese determined archaea's existence and found that it represented a whole new branch on the tree of life, all but a handful of unusually open-minded microbiologists scoffed and refused to acknowledge it. Now it is clear that archaea are at least as common as bacteria, much older than the rocks in our gardens, and capable of surviving anywhere. The archaea have tremendous implications for understanding the origins of life on Earth, for the discovery of life beyond Earth, and possibly for the future of biotechnology.

These are the main reasons I, long-time science journalist for the *USA Today* newspaper, wanted to tell this story, and how I ended up covered in slime in the Harlem Meer. What better place than the center of the world's fourth-largest city to look for new species of microbes that are adapting to and finding ways to make energy from the everyday by-products of excessive human habitation. New York City officials requested that if our diving team brushed against a decomposing body on the murky lake bottom we surreptitiously mark its location and notify them after we finished collecting samples. No need to spoil this occasion. A crowd has been watching the air bubbles of our tanks from the shore by the Dana Discovery Center.

Standing waist deep in the lake, diving mask pushed up over my head, I hand my samples to a volunteer who runs them back to a portable lab set up for the Central Park BioBlitz 2006. The event is marking the launch of the new E. O. Wilson Foundation. It is Friday, June 23, 2006. For me, it is the end of a long and wild adventure. For the scientists involved, it is a new beginning.

Edward O. Wilson is the world's most famous living naturalist. He popularized the term "biodiversity." Born June 10, 1929, in Birmingham, Alabama, Wilson has spent a lifetime studying insects, predominantly ants, teaching several new generations of naturalists and ecologists at Harvard about nature, and writing wonderful books on a wide range of topics. He published a famous tome on ants and won a Pulitzer Prize for his book *On Human Nature.* But perhaps he is best known for his controversial *Sociobiology.* Wilson argued that human behavior is influenced heavily by genetically driven instincts and not entirely by nurture or free will as some would have it. If this is true, we may trace some of our own behaviors, in particular, altruism and cooperation, to the genetically driven behavior of microorganisms.

E. O. Wilson was the 20th century's Charles Darwin, and he's still exploring, thinking, and writing as we speed along in the new millennium. He is a genuine celebrity scientist, the subject of several PBS documentaries. A new one on Wilson's life was being filmed at the Central Park event. Wilson and another famous naturalist, Peter Alden, organized the first BioBlitz at Walden Pond in 1998. A BioBlitz is essentially a species survey of a specific area or ecosystem. Hundreds of these events have been held worldwide over the past decade as a way to reconnect human beings to nature. But the Central Park BioBlitz 2006 is especially significant because it is the first one to emphasize the diversity of microorganisms. It turned out to be a huge success. More than 200 previously unknown species of microbes were discovered in the park. Not too surprisingly, microbes found in the sediment samples from the Harlem Meer appear similar in their metabolic signatures to microbes that have adapted to sewage sludge and toxic waste sites. Given enough time, microbes can become quite efficient at bioremediation. If we were to stop polluting the Harlem Meer today, it is quite possible that the microbes at work in the sediments and water would metabolize all of the pollutants and convert the lake back into pristine condition within a decade or two.

For this and many other reasons, Wilson has of late become enthralled with the microbial universe. With Wilson's influential pulpit and celebrity, I venture that microbes will be gaining in popularity too. It's about time. Microorganisms are absolutely essential to the cycle of life on Earth. They have dominated life on our planet for most of its 4.5-billion-year history. Both archaea and bacteria play tremendously underappreciated roles in Earth's geology, botany, biology, and in the composition of the atmosphere. Few people outside of the circle of scientists who study the ecology of microorganisms understand their full importance. But once one begins to witness the strange partnerships that archaea and bacteria have developed in acid mines, oil leaks, nuclear waste dumps, cancerous industrial waste waters, polluted city lakes, and, not the least, in our soils and oceans, one begins to marvel at their power. The ancestors of modern archaea and bacteria earned a very good living on Earth for more than 2 billion years before the planet became hospitable to life as we know it. Their continuing dominance in ordinary soil and water attests to their versatility. Microbes have had a very long time to hone their survival skills. There is not a single poison we can create that does not delight some several species of microbes. They appear to be able to adapt to anything. We cannot. We are entirely dependent upon microbes for our survival, but they do not depend upon us at all.

Three months before the new E. O. Wilson Foundation's launch in Central Park, Wilson gave the keynote address at the Explorers Club Annual Dinner. The honor is traditionally offered to the individual who blazed the newest trail of exploration. Wilson teased the formally attired audience of crazy men and wild women with the question, "What is left to explore?"

After a brief pause, Wilson answered in depth:

> Why, the biosphere of course, that razor-thin membrane of life plastered to the surface of Earth so thin it can't be seen edgewise from an orbiting space vehicle, yet is still the most complex entity we know in the universe. How well do we understand this part of the world? Proportionately not very much. We live on a little-known planet. Let me give you some examples. The best-studied animals are the birds, which have been

carefully collected by naturalists and explorers for centuries. Nevertheless, an average of three new species are added each year to the 10,000 already described by scientists. Comparable to them are the flowering plants: about 280,000 species known out of 320,000 or more estimated to exist. From there it goes steeply downhill. You'd think that the amphibians—that is, frogs, salamanders, and caecilians—would be comparable to the birds, but in fact they are still poorly explored: from 1985 to 2001, 1,530 new species were added to the 5,300 already found, an increase of over one-fourth, and with more new species pouring in.

When we next move to the invertebrates, what I like to call the little things that run the world, we get a fuller glimpse of the depth of our ignorance. Consider nematode worms, the almost microscopic wriggling creatures that teem as free-living forms and parasites everywhere, on the land and in the sea. They are the most abundant animals on Earth. Four out of every five animals on Earth is a nematode worm. If you were to make all of the solid matter on the surface of Earth invisible except for the nematode worms, you still could see its outline in nematode worms. About 16,000 species are known to science; the number estimated actually to exist by specialists is over 1.5 million. Almost certainly the world's ecosystems and our own lives depend on these little creatures, but we know absolutely nothing about the vast majority.

To continue: about 900,000 kinds of insects are known to science. I've just finished describing 340 new species of ants myself, for example, but the true global number could easily exceed 5 million. How many kinds of plants, animals, and microorganisms make up the biosphere? Somewhere between 1.5 and 1.8 million species have been discovered and given a Latinized scientific name. How many species actually exist? It is an amazing fact that we do not know to the nearest order of magnitude how many exist. It could be as low as 10 million or as high as 100 million or more. Those of us in biodiversity studies say that we have knowledge of only about 10 percent of the kinds of organisms on Earth.

The nematodes and insects and invertebrates all shrink in diversity before the bacteria and archaea, the dark matter of planet Earth. Roughly 6,000 species of bacteria are known. It's been recently estimated that a ton of fertile soil supports four million species of bacteria. We believe each one is exquisitely adapted to a particular niche, as a result of long periods of evolution. We don't know what those niches are. What we do know is that we depend on those organisms for our existence.

A search is on right now at least for the bacteria that live in the human mouth. The number of species adapted to that environment so far is 700. These bacteria are friendly; they appear to function as symbionts that keep disease-causing bacteria from invading. For those species, your mouth is a continent. They dwell on the mountain ridges of a tooth; they travel long distances into the deep valleys of your gums; they wash back and forth in the ocean tides of your saliva. I'm not suggesting that we give an Explorer's Club flag to a dentist. But you get the point. Every part of the world, including Central Park where a new kind of centipede was recently found, has new kinds of life awaiting discovery.

But—if none of this impresses you, would you like an entire new living planet for your delectation? The closest we may ever come is the world of the SLIMES—that's an acronym for Subterranean Lithoautotrophic Microbial Ecosystems—a vast array of bacteria, archaea, and microscopic fungi teeming below Earth's surface to depths of up to two miles or more, completely independent of life on the surface, living on energy from inorganic materials, possibly forming a greater mass than all of life on the surface. The SLIMES would likely go on existing if we were to burn everything on the surface to a crisp.

The morning of the foundation's launch, with the same sense of wonder, Wilson addressed BioBlitz volunteers, who had been crowded by a rain shower into a tiny room atop the Dana Discovery Center. Wilson told the crowd, "If I could do it all over again, and relive my vision in the 21st century, I would become a microbial ecologist. Ten billion bacteria live in a gram of ordinary soil, a mere pinch held between your thumb and forefinger. They represent thousands of species, perhaps tens of thousands, almost none of which are known to science. Into that world I would go with the aid of modern microscopy and molecular analysis."

This is the world into which we are headed now. Forget everything you have been told about life on Earth. Archaea are something completely different. As I mentioned, they have been found colonizing every conceivable environment, including all the places where life is not supposed to exist. Archaea swim in sulfurous hot springs, frolic in crude oil, laze in frigid methane seeps, burrow into bedrock up to *four miles* deep, colonize glaciers, drift in the stratosphere, and blossom beneath the sands of the planet's driest, most lifeless desert. (See Plate 1.) Whatever kills us makes them stronger. Archaea's diet includes iron, sulfur, carbon dioxide, hydrogen, ammonia, uranium, and all manner of toxic compounds. They produce methane, hydrogen sulfide gas, iron ore, and brimstone, to name just a few of their more common byproducts. They cooperate whenever possible with each other, as well as with bacteria and fungi. They possess metabolisms that can convert inorganic matter into organic molecules for their cells. Metal to meat. That is quite a skill.

When first discovered, archaea appeared alien-like in their predilection for oxygen-free habitats and for their incongruous abilities to obtain energy from raw inorganic compounds in the hottest, coldest, driest, darkest, most acidic and corrosive environments. The archaeon *Pyrolobus fumarii* is the hottest known living microbe on the planet. It was discovered in 1996 living at extreme pressures on a deep ocean hydrothermal vent. It dines on nitrates and sulfur and prefers temperatures of 236 degrees F.

Haven't we been taught since the days of Louis Pasteur to boil water to *kill* microorganisms? Boil it for archaea and they have an orgy.

Most other life forms cannot survive temperatures above 140 degrees F, let alone extremes of atmospheric pressure. We dehydrate and shrivel with the heat and either compress horribly to the point of becoming puree or we decompress—our brains and lungs swollen with fluids, our blood filled with air bubbles—with changes in pressure. Our cells begin to crystallize when left exposed to temperatures of 32 degrees F. Plant, animal, and human tissues require fluids with a neutral pH of about 7. Anything below 7 becomes more acidic, such as sulfuric acid, and begins dissolving our delicate cell membranes. Anything above 7 becomes more alkaline, such as soda-lime and potash, which is caustic and erodes plant, animal, and human tissues. Some archaea prefer their bath with a pH of 0, which melts flesh from bone in minutes. Others all around us thrive at the upper extreme pH of lithium hydroxide. Don't try this, but dropping lithium batteries in a bucket of water can generate a purple hydrogen gas fire. It's dandy for archaea.

Compared to the other planets and moons in our little solar system, Earth is the proverbial Garden of Eden for life as we know it, but only a thin slice of Earth. To paraphrase Carl Sagan who was paraphrasing astronomer Sir Fred Hoyle, drive a car about an hour toward the center of Earth and you'll melt into the mantle. Drive straight up for an hour and your coffee will freeze as the cup floats in space. Our domain of life exists only where we can walk on it, plow it, or swim in it.

Earth is blessed with moderate temperatures (for now); oceans of water that aren't too salty (until they evaporate); rainfall that doesn't burn our skin and kill the plants (except for acid rain falling from sulfurous skies); the right balance of nitrogen, oxygen, and carbon dioxide for plant and animal respiration (as long as we have phytoplankton in the oceans); and not too much methane (though melting tundra, cows, human sludge, coal mining, and natural gas refineries are a problem), ethane (petroleum refineries and biomass burning), or carbon monoxide (cars). The ozone layer (when there's no hole) helps minimize skin cancers caused by that big nearby star we call the sun. Earth is quite the remarkable garden, or at least has been for the last billion and a half years or so. For most of Earth's history, the planet has not been so friendly to our garden variety of life.

Scientists have defined the upper and lower limits of life by temperatures, atmospheric conditions, and pH. Everything within the surprisingly rare moderate limits is known as the habitable zone for life as we knew it. Plants and animals have the narrowest range of habitability and by far the shortest residency on the planet. Before Earth's carbon dioxide atmosphere was thoroughly poisoned by oxygen about 1.4 billion years ago, the Archean Empire, which included similarly "extreme" forms of bacteria, reigned supreme. Today, the "extremophiles" have been forced to the outer limits of our habitable zone by oxygen, but don't think for a second they couldn't rule the planet again. Keep dumping industrial poisons in the creeks. Pour used motor oil down the storm sewer. Dig that coal. Fill the aquifers with nuclear waste. Open the valves; let the greenhouse gases fly. It isn't hurting archaea and their bacterial co-conspirators one iota.

Something like archaea might be colonizing the soils of Mars, the volcanoes and oceans of Jupiter's moons, and maybe Titan's tundra and atmosphere. (See Plate 1.) Clouds, which are big chemical stews with plenty of available energy, could harbor life. Archaea could be packed inside hollow pockets of asteroids or microscopic veins of liquid water inside comets. Some extreme forms of archaea are especially suited to very cold temperatures. Most recently, colonies were found thriving inside cores of 10,000- to 15,000-year-old ice. Scientists estimate some of these cold-lovers undergo cell division once every few hundred to perhaps a couple of thousand years. The ability to survive under the most extreme conditions has made archaea the darlings of NASA's Division of Astrobiology, which is in charge of finding life beyond Earth in our little backwater region of the Milky Way Galaxy. If life is out there, NASA will eventually find it with the help of modern microbiology and the field's increasingly remarkable molecular tools.

About 90 species of archaea have been captured from their native habitats, isolated from samples of volcanic mud, deep ocean sediments, and hydrothermal fluids, and artfully cultured in the laboratory. But perhaps tens of thousands or millions more species are waiting to be discovered. The only archaeon to reach celebrity status so far is the headline grabbing *Methanococcus jannaschii*, which inhabits the photogenic hydrothermal vents on the ocean floor—the abyssal oases better known for the six-foot-tall tube worms swaying in the current and the white shrimp crowding around vent cracks. This was the first archaeon to have its genetic code fully sequenced and be declared on national TV to be a representative of the third domain of life, neither bacterial nor eukaryotic animal or plant life. This happened in 1996. The microbe had been selected for whole genome sequencing, and thus stardom, based entirely on its charismatic features and dramatic habitat. *Methanococcus* gets energy from hydrogen and carbon dioxide and produces methane gas as a by-product. No doubt it's a totally hip microbe living without oxygen at pressures that would crush your head and temperatures that would boil a cow.

But the truth is that *Methanococcus*'s hillbilly cousin, which lives in ordinary sewage sludge, was actually the first microbe determined to be an archaeon, and the first known member of the archaean or third domain of life. That happened in 1977, so most people probably don't know about this or have forgotten. Its fame was short-lived because few scientists at the time could grasp that any form of life other than bacteria (the first domain) and eukaryotes (the second domain)—which includes fungi, protozoa, plants, animals, and us—was possible. The proposal that the hillbilly methanogen represented a third form of life was immediately shouted down by leading scientists and then largely ignored for years. Unfortunately, that is how big science sometimes works. It's a dogma-eat-visionary world. Academic bullies usually get their way until the evidence for something entirely new becomes so great it can no longer be ignored, unless of course they think of it themselves. Evidence that archaea represented a new form of life was quite strong in 1977, and it was near bulletproof by 1980. That did not deter the critics. But by the time science had finally developed the technology to sequence *Methanococcus*, very few scientists debated archaea's rightful classification as a third major branch on the tree of life.

The more extreme forms of both archaea and bacteria were the first key links in the early cycle of life of Earth. For the first 3 billion years of Earth's history, extreme archaea and bacteria are thought to have represented the only known forms of life on the planet. As they evolved, they appear to have formed vast consortiums with each other. Fossil evidence dating back 3.5 billion years suggests archaea and bacteria built formations that covered large stretches of ocean bottom with odd columns, cones, and ridges. (See Plate 1.) These formations represent ancient "microbial mats." Living versions can be found today most commonly at deep ocean hydrothermal vents, in some shallow regions of the ocean, and on the bottoms of salt lakes and ponds. Microbial mats have captured broad scientific interest in recent years because of the weird metabolic partnerships taking place among many different species of microbes. The mats are essentially collectives, which become stronger and more efficient as a unit than as individual species on their own. This cooperative behavior and predilection for efficient energy use may have characterized much of early life on Earth. The innate drive for metabolic efficiency may be the single most important force that ultimately led to more complex forms of life, including plants, animals, and humans.

The early metabolisms of archaea and bacteria made it possible to convert inorganic matter, such as sulfur, iron, nickel, and uranium, and gasses, such as carbon dioxide, hydrogen, and nitrogen, into the organic molecules upon which all other life depends for energy. Scientists have created different descriptive terms for the various types of archaea and bacteria based on their environments and preferences for specific nutrients and temperature ranges. All microbes that produce methane are known as methanogens. Microbes that live in extremely salty seas and ponds are halophiles. Heat-lovers are thermophiles and the hottest of the heat-lovers are hyperthermophiles. Cold temperature microbes are psychrophiles, which have been found in recent years in the polar oceans and teeming in microscopic veins of water inside glaciers. The psychrophiles are the newest arrivals on the archaeal landscape, bringing with them huge implications for astrobiology.

After Carl Woese of the University of Illinois at Urbana-Champaign, determined the existence and phylogeny of archaea, Norman Pace of the University of Colorado, Boulder, learned how to go out into the environment and identify them based on specific segments of genetic code. Pace is regarded as the second key pioneer, after Woese, of archaea research in the United States. Pace's lab is probably the place to be today for young scientists interested in studying microbial ecology. Karl Stetter of the University of Regensburg in Germany is the other leading pioneer. Stetter is the individual who learned to sample and cultivate at least two-thirds of the 90 or so known archaea species. Everything that is known about archaea, and just about every-

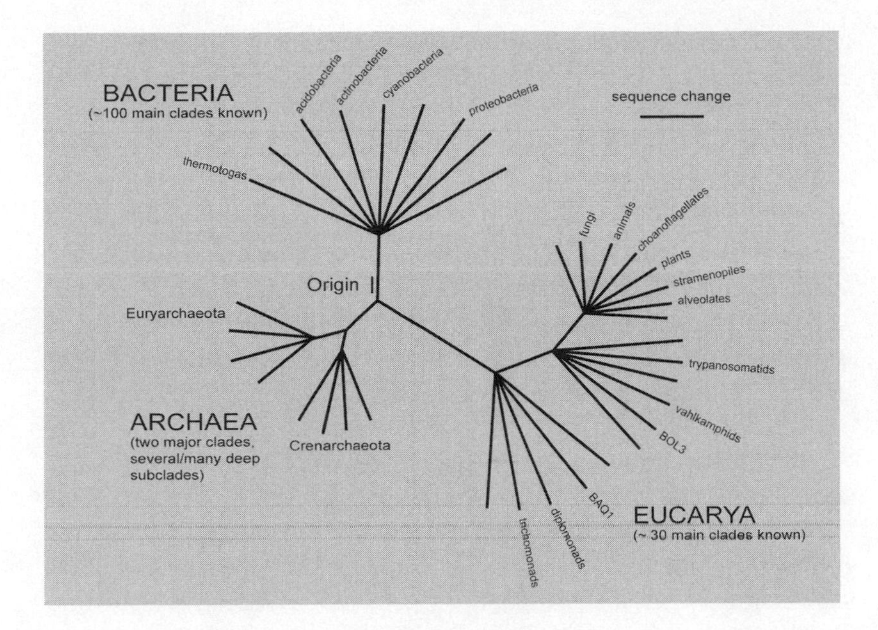

Norman Pace's tree of life. Courtesy of Norman Pace.

thing that is known about bacteria, is carried in the brains of Woese, Pace, and Stetter. They don't always agree with each other, but among them are all the keys to archaea's kingdom.

According to Pace, all known archaea are currently divided into two major branches:

- The **Crenarchaeota** include the archaea that live at the highest temperature extremes, such as *Pyrolobus fumarii* and *pyrodictium.* (See Plate 2.) Crenarchaeota grab electrons from hydrogen to reduce nitrates, sulfur, and other inorganic compounds to obtain energy and fix carbon dioxide. The Crenarchaeota include the new varieties discovered in soil and oceans living at moderate temperatures. The Crenarchaeota may be the most common type of archaea on Earth.
- The **Euryarchaeota** include methanogens, which live in anaerobic environments such as swamps and marshes, the rumen of cattle, sewage sludge, and termite guts. They use hydrogen as a source of electrons for reducing carbon dioxide. The by-product is methane, which is a highly flammable gas and the source of natural gas. Methane actually is odorless. The natural gas industry decided early on to add a

little hydrogen sulfide so we would know when gas was leaking. I suppose if they had added rose hips instead we'd all be blowing ourselves up in our kitchens. The Euryarchaeota also include halophiles, which are found in the Great Salt Lake in the United States and the Dead Sea. Halophiles maintain an osmotic balance with their salty surroundings by building up neutral "solute" concentrations of sugars, amino acids, and other compounds inside their cells. Putting halophiles in normal water destroys them. "Thermoacidophiles" are members that are found in the most highly acidic environments at temperatures not quite as extreme as those favored by the Crenarchaeota. Thermoacidophiles are found commonly in the hot springs at Yellowstone National Park.

The **Korarchaeota** are an extremely exciting new sub-branch—for now—of the Crenarchaeota. Stetter is expected to announce a major new discovery about the Korarchaeota in spring 2007. Korarchaeota's molecular signatures were first discovered in 1996 by Sue Barns, Pace, and colleagues at Obsidian Pool at Yellowstone National Park, but no one knew what they were. The discovery of the Korarchaeota resembles a true detective story. The molecular signatures were like having a genetic fingerprint of a suspect, but no positive ID and no matches found to any existing species (or suspects) in computer databases. The Korarchaeota's existence was based solely on genetic coding differences between known archaea. In 2000, Anna Louise Reysenbach of Portland State University discovered Korarchaeota fingerprints in association with a type of ancient bacteria called *Aquificales* at Calcite Springs, also at Yellowstone. Now the detectives had genetic fingerprints of the suspect at two locations. It turns out the temperature of both hot springs is around 83 degrees C and they prefer similar near-neutral pHs.

As detectives, knowing that the target had been spotted twice in the same type of environment, the scientists began searching for Korarchaeota's fingerprints at similar locations. Soon they found that Korarchaeota appeared to have a fondness for iron-rich fluids. A few attempts were made to culture the microbe using iron as a nutrient, but these failed. Reysenbach, Barns, and others began trying to grow cultures using the natural waters of the hot springs in which evidence of the microbe's presence had been found. This took some finesse. Microbes often depend on the metabolites of their bacterial and archaean neighbors for sustenance. This means cultures of the entire microbial community must be grown to nurture a single species. Growing up whole communities made it possible to finally isolate the Korarchaeota. In 2005, the microbial equivalent of the FBI got on the case. The Department of Energy's Joint Genome Institute (JGI) began applying its considerable technologies to sequence the whole genomes of all known Korarchaeota. This sub-branch may be one of the oldest lineages of life detected in nature so far, according to JGI director Eddy Rubin.

The **Nanoarchaeota** were first classified as a third major branch of archaea, but now they've been downgraded to a sub-branch of the Euryarchaeota based on work conducted at Pace's lab. This sub-branch, which was discovered by Stetter and proposed by him as a third kingdom, includes the very odd mini-hyperthermophile *Nanoarchaeum equitans.* So far, Nanoarchaeota have only a few official members. But 27 potential new candidates were discovered recently in one fell swoop in a scuba diving expedition at Yellowstone Lake. *Nanoarchaea* appear to be parasites, possessing a small and efficient genome for DNA repair and structural functions, but lacking genes for their own metabolisms. *Nanoarchaea* are very ancient, placed near the base of the tree of life. Their unusual lack of a metabolism would have made them especially suited to adapting almost anywhere in any environment that contained an archaeal host.

In terms of the origins of life, a number of scientists are convinced that it's silly to continue thinking life is limited to one tiny planet in an unimaginably vast universe with billions of galaxies. If extraterrestrial life exists, some scientists believe it probably shares genes with archaea and bacteria. The building blocks for DNA and RNA are present throughout the universe. Some of those giant gas nebulae out in space are full of sugars that form ribose—the backbone of RNA. Nature on this planet is extremely conservative with the genetic coding and cellular tools it uses to generate all the different forms of life with which we are familiar. There's no particular reason that the perfectly good system of DNA and RNA that shaped life on this planet should be limited to our biosphere. One theory of the origin of life suggests RNA coding is what actually gave primitive cellular structures the boost they needed to become life. With a universe full of sugar, perhaps early RNA worlds were generated and are evolving in their own unique ways in untold numbers of galaxies.

A growing number of scientists believe the discovery of archaea will have many significant impacts in our lifetimes. When *Methanococcus* was sequenced, scientists declared that archaea represent a "portal" into a new universe. Over the past 30 years scientists led by Woese, Pace, and Stetter have been sending molecular probes into microbial space and developing sensing technologies to explore it and bring samples back home to study. A remarkable synergy has been developing among geneticists, cell biologists, microbiologists, biotech entrepreneurs, geologists, and those men and women staring up at stars and wondering "Why not?" This interdisciplinary hybrid has been supercharged by a number of bold thinkers who were not afraid to challenge old assumptions, break ranks, and stake their careers on hunches—followed up of course with solid scientific evidence.

I had the opportunity to interview Carl Sagan on September 23, 1992, as he and Ann Druyan, co-author and wife, were promoting the book *Shadows of Forgotten Ancestors: A Search for Who We Are.* This was the first time I grasped that life beyond Earth was more probable than possible. Sagan adamantly expressed he was certain that life, especially microbial life, was "ubiquitous throughout the universe." He was more interested in "intelligent" life forms, but he believed the wide dispersal of microbes in the universe was a given, considering the odds. Intelligent life doesn't just appear. It must arise from something. As far as we know, all roads lead to microbes, which developed cooperative relationships as they evolved and may have eventually crossed a threshold into something more integrated and complex. Sagan explained that the universe possesses plenty of worlds on which life can experiment.

"In the observable Universe—containing as many as a hundred billion galaxies—perhaps a hundred solar systems are forming every second. In that multitude of worlds, many will be barren and desolate. Others may be lush and fertile, on which beings exquisitely adapted to their several circumstances are growing up, coming of age, and attempting to piece together *their* beginnings," Sagan wrote in the book.

Sagan was one of the pioneers of the field of astrobiology—the study of life beyond Earth. He kept good company. Though he was an astronomer, Sagan was influenced by a microbiologist—a bacterial geneticist. According to *The Lederberg Papers*, kept by the National Library of Medicine, astrobiology owes its launch to bacteriologist Joshua Lederberg. The notion of life on other worlds was purely fantasy and science fiction until October 4, 1957, when Lederberg saw *Sputnik* rise into orbit. He was in Australia at the time as a visiting professor at the University of Melbourne. From Australia, the fire of the Russian rocket was visible. Lederberg wasn't thinking about life on other planets, however. His imagination was fueled by concern that once we made contact with another planet or the Moon, if life existed there we might contaminate it or it might travel back on a spacecraft and contaminate us. At the time, Lederberg was 32 years old and would the share the Nobel Prize in Physiology or Medicine the next year "for his discoveries concerning genetic recombination and the organization of the genetic material of bacteria." After winning the Nobel, Lederberg was elected to the National Academy of Sciences and took advantage of his position to convey his concern. Had someone other than a Nobel Prize-winning bacteriologist brought up the topic of bugs in space, the Academy might not have listened. According to *The Lederberg Papers*: "In spring 1958, the Academy established the Space Science Board. . . . It urged that 'great care' be taken in sterilizing spacecraft before launch, and recommended a 'stringent' quarantine for samples returned from other planets until it could be determined that they were harmless." NASA adopted Lederberg's recommendations prior to its first flight to the Moon in 1969.

Michael Crichton's *Andromeda Strain* published in 1969 was inspired by Lederberg's work on the space board. The book is based on the premise of a satellite crashing to Earth, bringing with it a deadly space microbe. Lederberg wrote several letters to the publisher before the book was released demanding that Crichton's character be less obviously based on him.

Lederberg's concerns about space microbes were brought to light by an odd occurrence in the 1970s. Scientists had assumed that life cannot exist beyond Earth in part because cold and radiation would kill microorganisms. But, the common Earth bacteria, *Streptococcus mitis*, may have made a trip to the Moon and survived there—freeze-dried—for three years without nutrients, water, and radiation protection. The bacteria are thought to have made the trek to the Moon in 1967 aboard the *Surveyor III* spacecraft. The bacteria were discovered 31 months later when *Apollo 12* brought back *Surveyor*'s camera. The viable microbes were found inside a section of the camera. *Apollo 12* commander Pete Conrad commented in 1991: "I always thought the most significant thing that we ever found on the whole . . . Moon was that little bacteria who came back and lived and nobody ever said [anything] about it." Critics dismissed the discovery as lab contamination of the camera. Other experts aren't so sure. But based on what is known now about microorganisms, nothing should surprise us.

Lederberg coined the term "exobiology" to describe the study of possible life beyond Earth. In September 1962, he and Sagan officially launched the field with a joint paper published in the *Proceedings of the National Academy of Sciences* (*PNAS*). They wrote:

> The present capability of space flight gives timely urgency to discussions of problems of life on extraterrestrial bodies, especially Mars. . . . The low average temperatures, the low mean water vapor content, absence of extensive bodies of water of pure liquid water, low atmospheric pressure, absence of molecular oxygen and possible high ultraviolet radiation flux are undeniable constraints on Martian biology. Some terrestrial microorganisms survive in purported simulations of the average Martian environment. The permafrost hypothesis lends special importance to the distribution of localized geothermal activity—hot springs, fumaroles, volcanoes. Such hot spots should be accompanied by local outgassing and by higher water vapor pressures in the overlying atmosphere, making these locales much more favorable microenvironments for life.

Lederberg and Sagan were remarkably insightful. The hot spots they describe in their 1962 paper are all the same types of places where archaea would be found on Earth two decades later.

After meeting Carl Sagan, my next introduction to extreme forms of life also came in 1992, when Thomas Gold, a controversial astronomer and geophysicist at Cornell University, published a paper in *PNAS* entitled "Deep Hot Biosphere." Gold was the person responsible for bringing Sagan to Cornell in 1968. Sagan was very popular with the public for writing magazine articles and books about the cosmos. He had his own PBS TV show and was a frequent guest on *The Tonight Show* with Johnny Carson. Sagan died of cancer December 20, 1996, at age 62. Many scientists of the day criticized Sagan for pandering to the public and not conducting more research, but Gold believed in him. Gold was an even bolder maverick and truly maddening to scientists whose turfs he frequently trampled. Gold possessed brilliant intuition. He was wrong sometimes, but he had an uncanny track record of hunches about life and the universe that turned out to be true.

One morning on the science beat at the newspaper, I was drinking coffee with my feet on the desk and came across Gold's new article. I bolted upright and blurted "Hello! What's this? What does he mean, 'Deep Hot Biosphere'?" I called Cornell's news and information office and spoke with science writer Roger Segelken. He explained that Gold was famous for many ideas and discoveries, including the theory in 1967 that the source of mysterious but regularly pulsing radio waves detected in deep space were actually neutron stars emitting radio waves as they spin. Scientists had hoped that they were signals from a distant intelligent world. But as unlikely that might be, astronomers thought the idea of a star emitting pulsing radio waves was simply ridiculous.

"Gold's explanation was considered so implausible that he was not even allowed to defend it at a conference. However, the discovery of a pulsar in the Crab Nebula led to the theory's universal acceptance," according to Segelken.

Gold also coined the term "magnetosphere" to describe Earth's magnetic field. In 1955, before the *Apollo* Moon missions were planned, Gold cautioned that the Moon's surface would be covered in dust that could pose a problem for lunar modules and astronauts. He was among the 110 scientists who were given the first samples of lunar soil from the *Apollo 11* mission in 1969, which proved the surface to be powdery, though not as powdery as Gold predicted.

The one major hypothesis he appears to have gotten wrong was the "steady state" theory of the universe, developed in 1948 by Gold, Fred Hoyle, and Hermann Bondi. They proposed a constantly expanding but never changing universe. As the universe expands new matter forms to fill in the space. It was proposed as an alternative to the "Big Bang," a term Hoyle coined with sarcasm. Since about 1965, most of the evidence, particularly radio waves from deep space, has supported the Big Bang and discounted the steady state theory. More recently, astrophysicists have proposed a compromise "quasi-steady state theory."

Gold's theory about a deep hot biosphere was derided by geologists and most mainstream microbiologists. Gold argued that Earth's interior was teeming with microbes greater in mass than all of life combined on the surface. Evidence for Gold's case was already building among the handful of archaea researchers who saw it as a perfectly reasonable hypothesis. Today, this is stated as a matter of fact. Indeed, Darwin had thought along the same lines and said so long before anyone else:

> We may well affirm, that every part of the world is habitable! Whether lakes of brine, or those subterranean ones hidden beneath volcanic mountains, warm mineral springs, the wide expanse and depths of the ocean, the upper regions of the atmosphere, and even the surface of perpetual snow, all support organic beings.

Voyage of H.M.S. Beagle
Charles Darwin (1845)

Darwin possessed an uncanny intuition about the existence of extreme microorganisms. The "perpetual snow" he described was the permafrost, which is known today to contain many different types of microorganisms, including sulfur-loving species of archaea and bacteria and many methanogens. Scientists have found permafrost so thick with microbes that it appears to have been dusted with yellowish and sometimes pinkish particles.

Gold's Deep Hot Biosphere gave a big boost to scientists interested in the possibility of extraterrestrial life. In 1992 NASA officially launched its Division of Astrobiology. The timing was perfect for a confluence of ideas about extreme microbes. The archaea that Stetter had been discovering on hydrothermal vents and inside volcanic fissures possessed metabolisms that appeared ideally suited for other worlds. According to Gold, "As long as you think that life is possible only on planetary surfaces, the Earth is uniquely suitable. But when you talk about life deep below, the Earth is not unique at all. The deep, chemically supplied life may be common, not only in the solid bodies of the solar system, but throughout the universe."

Because it was such a strange concept at the time, my newspaper article on Gold's Deep Hot Biosphere theory made page one. A year later I spent an afternoon with Gold at his office at Cornell. It was delightfully cluttered with old books and scientific papers on hundreds of topics scattered about the floor. Gold, who looked somewhat rumpled, as an aging genius should, argued that due to the "fact" of the deep hot biosphere, "geologists have it all wrong about fossil fuels." This statement drove geologists batty. It remains contentious. The presence of organic molecules in oil is why geologists long ago concluded that oil is made from compressed organic matter. Hence, "fossil fuels." Oil is supposed to be the detritus of dinosaurs and plants and such, isn't it? Gold, who died June 22, 2004, maintained that the organic molecules present in oil and gas represented living organisms, which feed on hydrocarbons and produce methane. It remains unclear what role microbes are playing inside oil deposits, but archaea are no doubt major contributors to the cycle of life through their metabolic chemistries. Cooperatives of archaea and bacteria are certainly hard at work doing something in both oil and gas deposits.

No matter where life exists, it must possess a way to obtain energy. Life opportunistically develops different metabolisms to exploit any and all available sources of energy. The metabolism of a life form is defined as the sum of everything that happens in a cell with regard to using energy, taking in nutrients, and getting rid of what's left over. The two processes that constitute metabolism are catabolism and anabolism. During catabolism, organic or inorganic compounds are broken down and the energy contained in the compounds is transferred into the organism. During anabolism energy is processed to make new organic compounds that are honed into cellular components. Metabolisms are influenced by temperature, radiation, the presence or absence of oxygen, the amount of water, the abundance of organic and inorganic compounds, and, not the least, atmospheric and geochemistry.

Every living thing requires energy, but nothing under any circumstances in the physical universe as we know it can create new energy. All the energy that will ever exist was unleashed during the Big Bang. Perpetual motion machines and physical immortality are against the law. All life, as well as all of the stars in the universe, must ultimately surrender to the higher power of entropy. Everything must obey the laws of thermodynamics.

The first law states: Energy can neither be created nor destroyed.

The second law, as it pertains to biological organisms, demands that everything must die. The second law is defined as the dispersal of energy from a concentrated source. So, in a very real sense, we are all ice cubes melting in a warm room, or burning embers cooling and spreading our energy into space until we eventually turn to ash. While all individual life is held to the second law, it paradoxically evolved with a metabolism that is able to resist the law in the short term. The more efficient the metabolism, the greater the resistance and the more assured is long-term survival of life itself. The pioneers of the resistance to entropy are microorganisms. Hail Archaea!

Metabolism appears to have evolved first to obtain energy from inorganic and organic compounds made naturally from the chemical elements generated by the Big Bang. A chemical element is defined as "a substance that cannot be decomposed or transformed into other chemical substances by ordinary chemical processes." The elements are the same throughout the universe. Sulfur on Earth is the same as it is on Io, Jupiter's volcanic moon. Iron particles on Earth are the same as they are on Mars and in meteorites. Every element has a nucleus of protons and neutrons surrounded by a cloud of electrons. So far, 116 elements are known. The elements are grouped together based on their characteristics and behavior, and further arranged into chemical series, such as non-metals that include life's big favorites—carbon, hydrogen, oxygen, nitrogen, sulfur, and phosphorus. Life also is quite fond of the transition metals, such as iron, nickel, and zinc.

About 90% of the visible universe consists of hydrogen. All organic compounds contain hydrogen as well as many inorganic compounds essential to life. Archaea and the more extreme bacteria manipulate hydrogen as they go about the business of obtaining energy. Every element has different varieties of itself that are called isotopes or species. Isotopes of an element all have the same number of protons in their nuclei but different numbers of neutrons. Microbes display preferences for various isotopes of their favorite compounds. Armed with this knowledge, scientists can search for the chemical footprints of ancient microbes in isotopes of sulfur, for example. Sulfur isotopes with fewer neutrons are "lighter" than varieties with more neutrons. Life in general appears to prefer lighter isotopes of everything because less energy is needed to convert lighter isotopes into organic molecules. The sulfide by-product of sulfur metabolism ends up being a lighter isotope too. Microbial archaeologists recently found evidence of sulfur-loving microorganisms dating back 3.5 billion years, based on searching for the lighter isotopes. The oldest known chemical trace of life is about 3.8 billion years old.

Life as we know it in the universe is considered carbon based. Carbon is the foundation of all known organic chemical reactions on Earth, and, as far anyone can tell, everywhere else. But unlike most elements which arose from the Big Bang, carbon is incubated in the hearts of stars and released when a star is born along with all of the other ready-made ingredients for life. Carbon is an amazing element. As graphite, it is one of the softest known substances, and as diamond it is the hardest natural substance on Earth, maybe the whole universe. Generations to come will surely find diamonds anywhere in the solar system where volcanoes have been present at one time.

Science writer Ker Than for Space.com wrote recently about a mystery involving thick clouds of carbon enveloping a nearby star, Beta Pictoris. The star appears to be developing a planetary system at this moment. A study led by Aki Roberge of the NASA Goddard Space Flight Center in Maryland suggests the dense carbon mass surrounding Beta Pictoris might be aggregating into new planets. According to Goddard's Marc Kuchner, "If carbon-rich worlds are forming in Beta

Pictoris, they might be covered with tar and smog, with mountains made of giant diamonds. . . . Life on such a planet is not implausible, but it certainly would be exotic."

On Earth, at atmospheric pressure, carbon has no boiling point. Obviously it's not too affected by the blazing radiation of young Beta Pictoris either. One of carbon's chemical strengths is that it can form bonds with itself and so many other elements. It's involved in creating about 10 million different compounds. Carbon and oxygen form carbon dioxide, which was the primary source of carbon for early life on Earth. Carbon dioxide is still abundant in the atmospheres of other worlds. The thin Martian atmosphere is mostly carbon dioxide: 95.3%. The remainder is 2.7% nitrogen, 1.6% argon, 0.13% oxygen, and traces of water vapor, neon, krypton, and xenon. Hydrocarbons that we burn literally as fuel for our industrial complex are combinations of carbon and hydrogen. But when carbon combines with oxygen and hydrogen it makes fatty acids and the components of the cellular walls of all life forms. How beautiful is that?

Metabolism, however, is the most remarkable invention of all—at least to me.[2] It has evolved on one hand to take advantage of the free lunch provided by the sun via solar energy. Plants, algae, phytoplankton, and the even tinier picoplankton use solar power and carbon dioxide to make organic molecules for their cells and to store energy. Their by-products provide our modern atmosphere with oxygen, which our metabolisms require to obtain energy stored in plant cells and animal flesh. On the other hand, the roiling geochemical interior of Earth offers up "free" chemical energy for life at hydrothermal vents and at the mouths of volcanoes. The metabolism of many archaea and bacteria, which require no sunlight at all, are jump-started by the spontaneous chemical reactions that occur inside Earth and in hydrothermal vent fluids. Spontaneous chemical reactions trigger the metabolisms of archaea and bacteria to strip nutrients from inorganic matter and convert them into organic molecules. Similar spontaneous chemical reactions capable of providing energy to archaea and bacterial life are occurring in one form or another throughout the solar system and beyond.

As far as we know, microbial life on Earth has found a way to exploit every possible source of energy. This is why we find ancient lines of archaea and bacteria colonizing modern hazardous waste sites and thriving inside the cooling waters of nuclear energy plants.

Organisms that can obtain energy directly from sunlight or chemical sources, and from inorganic matter, are known as primary producers in the food chain. Plants are primary producers, and so are bacteria and archaea. Bacteria and archaea don't have to devour one another to survive; any chemical compound will do as their nutrient. In addition to obtaining energy directly from inorganic and organic matter, microbes cooperate to recycle the by-products of each other's metabolisms, thus making themselves even more efficient at resisting entropy. Microbes and their metabolisms have made it possible for consumers, such as humans, animals, and zooplankton in the oceans, to exist. All of the hard work necessary for extracting energy from a source element or compound has been accomplished for us by microbes. Indeed, our own metabolisms have evolved from those already found in microorganisms. As consumers, all we have to do to obtain energy is get our mouths around something that already has energy stored in it, such as an apple. Our cells and the more than 500 microbial species that live inside us then extract the energy, transfer it to make new organic molecules for our cells, store some energy, and burn the rest. We simply chew and swallow.

The biochemical reactions involved in metabolism run the gamut from relatively simple in the archaea to much more complex in eukaryotes. Either way, the essential purpose of metabolism is to restore some of the molecular disorder that occurs as energy disperses from a concentrated source due to the second law. According to the law, energy is always lost no matter what. That's why everything must die. We will never utilize 100% of the energy we obtain from the apple. But the more efficient we become at transferring energy and using it, the longer we get to use our bodies whether single-celled or multiplexed cooperatives. The microbial credo: Maximize efficiency. Waste little. Always resist the second law. Viva la resistance!

Metabolic chemistry for archaea and bacteria consists of reduction reactions and oxidation reactions—the so-called red-ox couples. These basic reactions consist of grabbing electrons from one compound or giving them away to another compound in order to obtain and use energy. In a reduction reaction, one or more electrons are taken from a compound as the microorganism acquires energy. A sulfur-loving archaea grabs electrons from hydrogen in water molecules and uses those electrons to convert sulfur into hydrogen sulfide gas. In an oxidation reaction one or more electrons are given up. An organism is always reducing something and oxidizing something else, hence the term red-ox couple. Energy is transferred from one form to another with the help of the "reducing agent" or "oxidizing agent" as the catalyst. The agents vary but carbon dioxide and hydrogen are biggies for microbes.

For life as we know it, nitrogen is a very significant player. It makes up 78% of Earth's atmosphere. Nitrogen also is a key component of the nucleic acids that make up DNA and RNA, which form amino acids, which form proteins. Humans and animals get nitrogen by breathing air. But under our feet in the soil, utilizing nitrogen is another story. All plants need nitrogen, of course, but they cannot absorb it from the air. Nitrogen must settle into the soil where soil bacteria absorb it and convert it chemically into nitrates that plant root systems are then able to use as fertilizer. No soil bacteria, no plants.

Roy Cullimore is a Canadian microbiologist who began his career as an agricultural microbiologist. Today he studies the cooperative relationships of microbes. He views Earth's surface as one big round petri dish upon which soil is merely a culture medium for microbes. Soil bacteria are specialized at "fixing" nitrogen. A number of species are involved at breaking it down and converting it into ammonia, nitrate, and nitrogen dioxide. Martinus Beijerinck worked out the concept of nitrogen fixation in the late 1800s.

When plants and animals die, the bacteria that specialize in extracting the remaining energy of the newly departed release ammonia. Since microbes waste next to nothing, the energy contained in the ammonia is extracted by other microbes. This process, which converts ammonia into nitrates, is called nitrification. Sergei Nikolaievich Winogradsky discovered soil nitrification in the late 1890s. Winogradsky, who was originally a specialist in sulfur bacteria, invented the "cycle of life" concept. He found that two different species of bacteria are actually involved in converting atmospheric nitrogen into organic compounds useful for plants. One species oxidizes ammonia into nitrite—meaning the nitrogen compound loses its electrons to oxygen—while the second species oxidizes nitrite into nitrate, which finally makes energy available to plants. Very recently, scientists have found that the archaea living in soil have a hand in this nitrogen-fixing and nitrification business, but they have no idea yet to what extent. This is a new chapter in archaea's history still waiting to be written.

This soil surprise was the product of an experiment conducted by an international team of researchers from Norway, Germany, the United Kingdom, and the United States. The team was initially looking for any microbial genes associated with the ammonia-oxidizing bacteria—the nitrifiers. They analyzed soil samples from 12 pristine and agricultural regions in three climate zones. They wondered what differences might exist in concentrations of the microbes and the gene coding between microbes found in pristine soils and in heavily farmed soils. The study, led by Christa Schleper at the University of Bergen, Norway, narrowed the focus to a specific gene for a key ammonia-oxidizing enzyme. The same gene had been detected recently in a species of ocean archaea belonging to the Crenarchaeota. The ocean archaea also were a big surprise. These Crenarchaeota were found in the upper sunlit zone of the ocean where most of the carbon dioxide of the atmosphere is absorbed and 50% of the global oxygen supply is generated. With this in mind, Stephan C. Schuster of Penn State University, and his colleagues, measured the quantities of bacteria and archaea genes expressed (meaning they are turned on) in the complex mixtures of soil organisms.

The team found that nitrifying archaea were up to 3,000 times more abundant than nitrifying bacteria. No one had any idea. They also found archaea at work at deeper depths in the soil than the bacteria. What does this mean? It's too soon to know, but it certainly means it's time to go back to the drawing board as far as the nitrogen cycle of the planet's soils are concerned.

"It might mean that archaea can oxidize ammonia at least with less oxygen and probably also with less ammonia, but we don't know for sure. Our data clearly say that the archaea are more versatile in their lifestyle than bacteria," Schuster said.

This strongly suggests archaea are the most common microorganisms in soil. If this is true, it really messes up what scientists thought they knew for the past century about the cycle of life. One might ask what survival advantage might exist in soil possessing so many nitrifying archaea? Considering Earth's atmosphere went through about a billion-year-long transition from a carbon dioxide atmosphere to one saturated with nitrogen and oxygen, plants could have gotten started with much lower levels of oxygen than we have today. And if oxygen levels should begin to fall again—say due to global warming—soil nitrification for plant growth could continue for a long period with so many archaea on the job.

How can archaea be so versatile? It turns out many Crenarchaeota contain a special molecule named called "crenarchaeol" that appears to allow their cell membrane to adapt to different temperatures. This molecule would explain how the Crenarchaeota are able to resist the highest temperatures of any known organism while evolving species adapted to moderate temperatures in soil and the oceans and even to cold temperatures. Crenarchaeol has recently been found in the psychrophiles. What a useful tool to have in one's evolutionary bag of tricks. Imagine what special chemicals archaea might develop to adapt to changes of extreme cold and heat should they reside inside comets or on asteroids. The incredible versatility of these organisms suggests to me that no geographic or planetary boundary should limit archaea's range.

The evolution of early life is tied closely to the metabolism of iron, which is tied to the spontaneous geochemistry of Earth—as well as the geochemistry of the solar system. Species of bacteria and archaea that specialize in iron may have developed one of the first metabolisms on this planet. This form of metabolism might be at work right now inside Martian soils, or at a few hidden geologically active hot spots, and even in Mars's polar ice caps. Scientists announced as late as December 8, 2006, that they had found evidence for liquid water and fresh sediments on Mars. Why would archaea not be present? All of the ingredients are present, including a thin carbon dioxide–filled atmosphere similar, perhaps, to the one on early Earth.

The "Red Planet" gets its name of course because of the high iron content at its surface, which eroded from rocks. Early Earth's surface might have looked pretty red too, but our surface iron would have been dissolved as soon as we developed the global ocean. What a rich milieu for growing ocean microbes that would have been! Deep inside Earth, iron ore has been manufactured over the eons by iron-working microbes, perhaps descendants of the early iron-feeders in the first ocean. Some of Earth's magnetism might also be "biogenic" in origin. Magnetic iron particles, possibly by-products of microbes, have been incorporated into animals and insects and used for navigation. Geology and biology on Earth are much more closely linked than anyone imagined.

"The Earth's geochemistry has changed substantially over time, in large part due to the influence of the metabolic 'inventions' by bacteria. A classic example of this is the evolution of photosystem II [a key component of the machinery of photosynthesis], which enabled cells to evolve molecular oxygen from water and thereby oxidize the Earth. Prior to this invention, however, for millions and perhaps billions of years, microbial life had to subsist anaerobically. How did cells cope? What electron acceptors and electron donors did they use for growth? It is increasingly accepted that ferric iron minerals may have been the most abundant and the most important terminal electron acceptors for ancient cellular respiration," according to Diane Newman at the California Institute of Technology.

Earth most likely started out with primarily a carbon dioxide atmosphere and little to no oxygen, giving anaerobic archaea and bacteria a field day for about 2 billion to 2.5 billion years. Starting around 2.4 billion years ago, oxygen levels in the atmosphere began rising as microbes blanketed Earth and consumed increasingly greater amounts of carbon dioxide. Oxygen was possibly freed from the crust and was also produced as the by-product of wildly abundant anaerobic metabolisms. The Achilles heel of anaerobic microbes obviously is oxygen. Quite possibly, the Archean Empire was brought down in a Malthusian catastrophe, poisoned by its own runaway success.

Little of this information is generally known by people outside of the relatively small group of scientists who study archaea and extreme forms of bacteria. As a reporter, information about archaea dribbled through the mainstream scientific press to me only in fits and starts. In the 1990s, the focus on microbes at the newspaper level was mostly on sequencing their genes and whatever that meant in terms of making money for biotechnology companies. After Thomas Gold's Deep Hot Biosphere in 1992, the next big headline didn't come until *Methanococcus*'s genome was sequenced in 1996. Another two years passed before anything newsworthy about life in the outer limits crossed my desk again. In 1998, the Diversa Corporation and Karl Stetter, who is a co-founder of the company, worked out the entire genetic code for a bacterium called *Aquifex aeolicus*, which Stetter had discovered in another hostile environment where life was not supposed to exist. I had become familiar with Diversa by that time, but as a company they had not made much news. *Aquifex* was a cool story because the organism was found thriving at 200 degrees F in hydrothermal vents on the ocean floor. Stetter's long-time obsession has been finding organisms at the highest temperatures—at or above boiling water. He coined the word *hyperthermophile* in the 1980s to describe the highest temperature organisms, whether archaea or bacteria.

For sustenance, *Aquifex* requires only hydrogen, carbon dioxide, mineral salts, and a tiny fraction of oxygen—about 28,000 times less than the content of our atmosphere. The by-product of its respiration is water. Based on comparisons of *Aquifex*'s genome with those of other fully sequenced microbes—more than 15 by 1998—*Aquifex* appeared to be a very early form of bacteria. My article on *Aquifex* carried a headline that read, "Microbe from ocean floor may resemble life in space." I wasn't sure if that was true, but who would argue? My story emphasized that *Aquifex* (which means water-maker) was one of the most heat-resistant life forms on Earth, thus providing a clearer picture of what life might resemble if it exists in extreme environments beyond Earth. But what was actually most noteworthy about *Aquifex* is that somewhere back in the recesses of time it had acquired 16 genes otherwise unique to archaea. This sharing of genes with archaea was downright strange. Woese and Pace, who had developed a very good phylogenetic tree of microbes by 1998, discovered that this "horizontal gene transfer" between archaea and bacteria really screwed up sorting out the roots they were hoping to find at the base of their tree of life. The genome studies of microbes were suggesting that archaea and bacteria had been swapping genes in the very early stages of life on Earth. In a volatile world with newly evolving metabolisms, this might be a handy survival mechanism. Diversa's study was among the first to demonstrate the phenomenon of horizontal gene transfer.

I began paying closer attention. The unusual metabolisms of extreme microbes had attracted the interest of two key biotechnology pioneers. Jay Short is the individual who created the E.O. Wilson Foundation and who previously co-founded the Diversa Corporation in 1994. J. Craig Venter recently founded the private company Synthetic Genomics Inc., and in 1992 launched The Institute for Genomic Research (TIGR). The same business deal that allowed Venter to found TIGR led to the birth of Diversa. With funding from the Department of Energy (DOE), TIGR performed the genetic sequencing that turned *Methanococcus* into a celebrity.

Short and Venter were among a small number of scientists who early on saw the potential for using extreme microbes to generate biofuels; revolutionize the chemical, pharmaceutical, and agriculture industries; and possibly clean up hazardous waste. They pursued somewhat parallel paths, but each had his own goals and each developed different but complimentary methods for getting what he wanted from microorganisms. Venter went after whole genomes and genetic sequence data. Short went for the amino acids and proteins made by microbes.

In the early to mid-1990s, the microbial world was dividing into two scientific hemispheres. In one, Woese, Pace, and their academic followers pursued the phylogeny of microorganisms. In the other, Short and Venter explored the exploitation of archaea and bacteria to make money while (honestly) wanting to help save the human race from its recklessness and horribly inefficient ways of using energy. Stetter kept his feet planted in both hemispheres. Biotechnology and microbes held great promise for helping the federal government save hundreds of billions of dollars in costs related to cleaning up hazardous waste sites.

If you were to create a phylogenetic tree, so to speak, of the key researchers involved with archaea and extreme bacteria, you would find that Short, Stetter, and Venter are all closely related as co-founders of Diversa, though Venter was a co-founder only in spirit—and stock options. A group of venture capitalists (VCs) originally set Venter up with enough money to operate TIGR for about 10 years. To earn their investment back, the VCs created three for-profit companies to develop products from the gene sequencing data generated at TIGR. One of the companies was Human Genome Sciences, the second was Industrial Genome Sciences (IGS), and the third was Plant Genome Sciences. Mel Simon, an early pioneer of biotechnology based in San Diego, was instrumental in establishing Industrial Genome Sciences with Venter and the venture capitalist Wally Steinberg. Simon was attracted to Stetter's menagerie of archaea and invited him to join the company. But very soon after IGS was founded, Simon and Venter and the other co-founders, now including Stetter, decided to give the VCs their money back and go start their own company. The VCs let the IGS folks out of the deal, but offered to give the same money back to form an independent company, which was called Diversa. Venter's stock options in IGS were transferred to Diversa, technically making him a co-founder. Short and Venter both developed very close relationships with the DOE, which began funding microbial research in 1994 (not coincidentally) through its Microbial Genome Initiative. The DOE program's primary goal was and still is to develop biofuels and tools for bioremediation. The Genomes to Life program launched in 2001 funds about $135 million a year in grants to academic scientists—actually a fraction of DOE's $3 billion a year Office of Science budget. But NASA and the National Science Foundation add a few more millions to funding of microbial research for academics. I've heard many scientists state over the past few years that if the federal government were to start funding microbial research on the scale of what was invested in the atomic bomb project, it could revolutionize our energy, chemical, and agriculture industries within a decade. I agree. But I doubt it will happen until it becomes absolutely necessary.

A tremendous amount of knowledge about the behavior of microorganisms has been gained since the discovery of archaea in 1977.[3,4] Until 2003, archaea was entirely synonymous with "extreme." Though similarly extreme behaviors exist in bacteria and archaea, it wasn't until archaea's discovery that scientists began to search in earnest for bacteria with the same habits. So it is the discovery of archaea that triggered a modern revolution in microbiology and brought us to the threshold where we stand today. And so I am focusing primarily on archaea, even though bacteria are certainly as versatile and important.

In full disclosure, archaea's importance to science was not the only reason I became interested in the topic. As it happened, Diversa was the only group—whether academic or private—that was developing the technology to extract raw microbial DNA from environmental samples and use that DNA to find active proteins. To obtain their samples, Diversa scientists were engaging in expeditions to remote regions all over the world.

Some reporters will do almost anything to get a great news story. Others will do almost anything to get out of the newsroom. That would be me, waiting for the next phone call from someone with an adventure to offer and any angle that could qualify as science. As a science reporter at *USA Today* for 16 years, I had been able to make occasional forays into the microbial universe. But in 2003, after a series of events that steadily built my curiosity about archaea and all manner of extreme microbes, I plunged full-time into this exotic and unknown realm. My chief guide for the scientific and adventure journey has been Short, who launched the E. O. Wilson Foundation to promote biological diversity, in general, with Wilson and, in particular, to promote the importance of microorganisms to the public and to industry. Short has a long history with microbes. From 1994 to 2005, Short and his scientific team at Diversa developed many of the fundamental technologies used by scientists everywhere today to discover microorganisms in all types of environments and manipulate their genes and proteins. In the beginning, even Short's own team thought what he wanted to do with microbes was impossible. But the company's technology quickly attracted attention in the scientific community. Short and his team developed multiple successful collaborations with scientists at universities and with the DOE/JGI and the National Park Service. It was unusual to see a private company as involved in pure science and research development as Diversa. And it was unusual to find a team of scientists in the private sector as respected by academic and government scientists as Short's group.

While at Diversa, Short demonstrated that it is possible to exploit microorganisms to reduce the need for harsh and expensive chemicals used commonly in many industrial processes. Now he hopes to spread that knowledge and try to convince all industries to become more eco-friendly through the power of microbes. It is an ambitious goal.

I had known Venter since 1991 when he was still a scientist at the National Institutes of Health working on a new way to sequence human genes. Like many reporters, I had followed Venter's work with acute interest over the years. In 2003, I saw a kind of "perfect storm" developing in the area of extreme microbes and biotechnology. Venter had finished sequencing the human genome and had just established the new non-profit J. Craig Venter Institute to focus on the human genome, the environment, and synthetic genomes. Venter was moving into Diversa's long-held microbial territory, and he was doing it with money and style. He bought a 90-foot sailing yacht to begin sampling microbes in the world's oceans. The DOE was funding research at both Diversa and the Venter Institute to speed their goals. Archaea and bacterial research were becoming mainstream and totally cutting edge. It was hot, as my editors used to say.

I was aching to get a closer look at Diversa's field activities, but so far I had not been invited on a date with their scientists. Finally, in mid-April 2003 I was sitting at my desk, when the phone rang. Diversa's public relations person was on the line. Would I be interested in accompanying a Diversa scientist named Eric Mathur to Costa Rica in two weeks, for about four days to collect microbes from the jungle?

Ka Ching!

This became the beginning of an odyssey that would lead to many more expeditions and culminate in June 2006 with the announcement of the birth of the E. O. Wilson Foundation in New York City's Central Park. No one in this story had any idea at the time how much their lives would change in three short years.

2
BUG HUNT

Most children have a bug period,
and I never grew out of mine.
Edward O. Wilson, *Naturalist*

Dennis Kelly, my editor at *USA Today*, was making me nervous. He actually appeared interested in sending me to Costa Rica on Diversa's expedition. He wanted to know more about Diversa's bioprospecting of microbes and the company's new technologies for developing biotech products. What exactly was Diversa planning to do in the jungle? Dennis and I were standing by his desk in the newsroom, talking in hushed tones because some reporters in my section considered these assignments to be "company paid vacations." I murmured that Diversa would collect samples of microbes from different ecosystems in the rainforest, and they wanted to showcase their technology and royalty-sharing arrangement with Costa Rica. Diversa had been singled out by the United Nations and the White House as an example of how bioprospecting could be done ethically and profitably for companies and the countries that own the natural resources.

The Diversa team would consist of Eric Mathur, Leif Christoffersen, Steven Briggs, and their media handler. Mathur was their expedition guy and senior director, Christoffersen was their biodiversity coordinator, and Briggs had joined Diversa just weeks earlier as their chief division scientist. Briggs previously had led the effort to sequence the genome of rice at the Torrey Mesa Research Institute, which was a significant achievement. As a member of the National Academy of Sciences, the scientific community took him seriously. I did not know Briggs had *founded* the Torrey Mesa Research Institute for Novartis, which would later merge its agricultural division with AstraZeneca's to form the world's largest seed and agriculture company named Syngenta Corp. Syngenta was heavily involved in developing genetically modified crops. I also did not know at the time that Briggs is an expert scuba diver and drives a Maserati Coupe' Cambiocorsa with paddle shifters. Briggs ended up at Diversa after it traded some stock to Syngenta for the Torrey Mesa Research Institute. Who wants to be an instant multimillionaire?

My contact over the years with Diversa had been limited to phone interviews with Diversa's CEO Jay Short. I had never met Mathur or Christoffersen and didn't know much about them except that Mathur was said to be "charismatic and colorful" and that Christoffersen was earnest and fluent in Spanish. Both got paid for collecting microbial samples from extreme environments around the world and for developing royalty-sharing agreements with the countries in which they go bug hunting.

Glancing over my shoulder, I ensured no one was eavesdropping and continued describing the plan to Dennis. Diversa would meet a team of Costa Rican collaborators from INBio—the Instituto Nacional de Biodiversidad—in San Jose. INBio is a private, non-profit research and biodiversity management center. A quick Google search revealed INBio was established in 1989 to catalog and study Costa Rica's abundant biological diversity, and to find sustainable ways to make money from it. INBio collaborates with government institutions, universities, biotech and pharmaceutical companies, and public and private organizations in and out of the country. Diversa had updated INBio's lab equipment and trained its staff to extract DNA from bugs, swamp water, and dirt.

After a half-day meeting at INBio, the group would drive from San Jose to the southeast Caribbean coast to collect samples at the Gandoca-Manzanillo Wildlife Refuge. Once in the jungle, the group would spend three or four days collecting insects, sediments, and water samples from different parts of the rainforest. The microbial DNA would be shipped to Diversa for gene expression screening, sequencing, and tinkering. Dennis glanced over my shoulder at someone who was staring at us, lowered his voice and asked, "What do you mean 'tinkering'?"

"I don't really understand it yet, but they have these processes Short invented, which he calls 'environmental library discovery and non-random directed evolution.' First they express all the genes representing microbes in a sample. Then they do automated culturing of all the microbes and find the proteins made by the genes. Then they break down the proteins into their amino acid sequences and somehow recombine the sequences at random to create new proteins or enzymes. Short told me they just started developing a new commercial enzyme based on 'evolving' the amino acids of a microbe their team found in a hot spring on an expedition to Kamchatka in Siberia in 2000."

Amino acids are created from the genetic code. Each of the 20 amino acids has its own unique sequence of code. Diversa's "directed evolution" involved generating permutations of amino acid sequences. The new enzyme they were developing had a specific desired pH, high-temperature resistance, and unpredictable abilities to bleach wood pulp whiter than is possible using dozens of chemicals.

"Why do they want DNA from microbes living in the intestines of insects?" he persisted. I explained Diversa wanted microbes from the intestinal tracts of rainforest insects because (a) rainforests are incredibly diverse and dynamic environments, which places all kinds of interesting selection pressures on microbial genes, and (b) intestinal microbial fauna are specialized to break down foods that are difficult for their symbiotic hosts to digest. To convert indigestible flora into energy the microbes must manufacture powerful enzymes to dissolve and catalyze them.

Termites were at the top of Mathur's wish list for the Costa Rica trip. Termites eat wood. The microbial enzymes inside termite bellies are very good at digesting wood. Termite hindguts are home to 100 to 250 species of microbes, including cooperative communities of anaerobic archaea and anaerobic and aerobic bacteria. The mix of intestinal enzymes is so efficient that it completely converts wood in the termite's gut into sugars in about 24 hours. Imagine pouring tons of that stuff on many more tons of wood pulp. A few more conversions and you have ethanol. The by-products of the microbes' metabolisms are mainly hydrogen and methane. Hydrogen is so light it's very hard to trap. But microbial methane could be used as feedstock for natural gas refineries. Two bangs for one buck.

Dennis folded his arms and glanced out from our giant glass house in Tyson's Corner, Virginia, at the Dulles Toll Road. "Okay," he said. "Let's do it. Let's do a full cover story with an inside jump, photos, and sidebars."

I began compiling background information for the Costa Rica assignment. Diversa had been producing good research, publishing in reputable journals, and patenting like wild for about six years at this point.[1] The company was popular within the scientific community but did not attract the attention of Wall Street until 2000 when Diversa went public.

"It was impossible to get Wall Street interested in microbial biotechnology before that point. No one besides us and a handful of people in the scientific community believed that the genomes of uncultured microbes would be valuable. Up to the point when we went public, it had been bloody impossible to get people to look at us and understand what we were doing. It took us six years of hard work. We really started the process at Stratagene, where Eric and I were before we co-founded Diversa," Short told me before the trip.

Wall Street is a fickle lover. It bathes you in adoration and showers you with money early in the courtship. It then dumps you at the first hint you aren't putting out as quickly as it hoped—and the analysts hyped. When Diversa went public as a company, the stock was offered at \$24 a share ending at \$57 on that first day, but within a couple of weeks it hit an unbelievable \$140 a share. Venter wisely sold his stock in Diversa while it was at its peak. The stock dropped back into the 20s within the next six months after the inevitable 2000 bubble burst, ending up in the \$10 region throughout 2005 and 2006.

On the flight from Washington, D.C., to San Jose, Costa Rica, I read the background material I compiled and the files on Diversa's product pipeline and biodiversity agreements. As a company basing its business on microbes, Diversa had plenty of raw material to work with. Scientists estimate more than a billion species of microbes currently colonize Earth. They're everywhere—on your skin, in your nose, in your hair, in your gut, in water, soil, and air—all driving the chemistry of life, affecting global climate and cycling critical elements such as carbon and nitrogen to maintain our atmosphere and keep our habitable zone intact. The current microbial industrial complex generates at least half the oxygen we breathe. The Department of Energy (DOE) and biotechnology industry believe harnessing the genes of archaea and bacteria, especially the stranger ones, will rescue the habitable zone of Earth from the mess humans have made of it in a brief 200 years.

To date, scientists say fewer than 6,000 of the estimated 1 billion species of microbes on Earth have been identified. This is because the vast majority of microorganisms cannot be grown in a laboratory petri dish. Until Carl Woese developed a method in the 1970s of identifying microorganisms based on a little piece of RNA coding, no one had any idea how to find them.

The first technology for identifying microbes was the microscope, which Antonie van Leeuwenhoek developed in the mid-17th century to have a better look at the hidden universe of "animalcules." Leeuwenhoek also appears to have been the first scientist to conduct a microbial sampling expedition and write of his discoveries. He went sailing on Berkel Meer near his home and filled a little bottle with water. After putting some drops of the lake water on a slide he peered into his microscope and then wrote on September 7, 1674, in a letter to the Royal Society of London:

> Passing just lately over this lake . . . and examining this water next day, I found floating therein divers earthy particles, and some green streaks, spirally wound serpent-wise, and orderly arranged, after the manner of the copper or tin worms, which distillers use to cool their liquors as they distil over. The whole circumference of each of these streaks was about the thickness of a hair of one's head . . . all consisted of very small green globules joined together: and there were very many small green globules as well.

The scientific community at the time believed that microorganisms arose spontaneously from inanimate matter. Nearly a hundred years after Leeuwenhoek, Lazzaro Spallanzani, an Italian naturalist, designed a clever experiment to disprove this widely held notion. (Ironically, today's scientists are back to trying to prove microorganisms actually did arise from inanimate matter with only a couple of evolutionary steps in between!) The first microbial culture on record was made in 1795, though it wasn't appreciated for what it was. Spallanzani began his experiment by pouring some broth into two vessels. One vessel he left open to the air. He boiled the broth in the other container and sealed it to keep any microorganisms from floating in the air. Broth in the open vessel soon was teeming with microorganisms, but the boiled broth in the sealed container remained sterile. For Spallanzani this was proof that microbes were not spontaneously generated. Otherwise, microbes should have appeared in the covered container. He then isolated a single bacterium and witnessed its division into two bacteria. This was a true beginning for microbiology, but mainstream science at the time was not about to fall for such nonsense. It was not until the 1860s that Louis Pasteur finally convinced the world that microbes are not spontaneously generated. He also succeeded in promoting the germ theory of disease.

The next big technological revolution came in 1882 when the German microbiologist Robert Koch got down to the business of cultivating microorganisms in the lab and discovered the bacterium responsible for tuberculosis. To grow his bacterium, Koch's wife developed a nutritional agar upon which microorganisms would grow into colonies on plates. Agar is a kind of seaweed Jell-O. Koch named the plates "petri" dishes, for his assistant Julius Richard Petri, who invented them.

Culturing has remained the standard method of identifying microbes until quite recently. It has always been a tricky business to figure out what specific nutrients will stimulate microbes to start a colony. Scientists estimate that only 1% of microbes will grow on standard agar. Over the years, scientists have experimented with adding different nutrients to agar, but it has been an art that produced little success.

Culturing of environmental samples of lake water or soil yields only about 1% of the species that are present in the sample. The one that grows is the species that is often most amenable to the culture medium. These are regarded as the "microbial weeds." This inability to culture microbes caused scientists to vastly underestimate the number of microbial species on Earth. Today it is clear that any given handful of dirt or vial of swamp water is teeming with thousands of unknown species. Microbiologists estimate the number of species in a handful-sized environmental sample at around 5,000, making the microbial biosphere of Earth truly uncharted territory. Microbes constitute at least half and possibly 60% of the mass of all living things on our planet—and inside the planet as Gold predicted in 1990 and Darwin before him in 1848.

So far, fewer than 6,000 species of microbes have been identified and fully characterized using cultures. About 90 species of these 6,000 are archaea, and of these about 60 have been cultured by Karl Stetter, who developed extraordinary measures to grow these various methane-loving, salt-loving, sulfur-loving, heat-thriving, and pressure-dwelling oddballs in the lab. Culturing is rapidly becoming passé for microbial ecology and species surveys. It is the wrong tool for the job. Obtaining a little bit of DNA from microbes and applying polymerase chain reaction (PCR)—genetic fingerprinting—to amplify the genetic bits obviates the need to culture. This permits scientists to determine the true number of species in a sample and learn which ones are new.

Any number of milestones could be chosen as a starting point for a little history on the development of genetic tools that allow discovery of entire microbial communities in samples from the environment. But we have to consider that two revolutions were taking place in parallel. One was the study of DNA and RNA and their components called nucleotides that sprang from the discovery of the double helix structure of DNA in 1953 by James Watson and Francis Crick. The other was the development of microbial ecology to unmask the identities and biochemistries of all the species in a sample. Microbial ecology is interested in the identities and dynamics of the organisms resistant to culturing.

First, on the DNA/RNA side, the discovery of the double helix helped Sidney Brenner and Francis Crick determine in 1961 that life is a three-letter word—as Sagan put it—determined by strings of three nucleotide bases of RNA that instruct a cell to manufacture individual amino acids. By 1965, scientists learned which nucleotide bases represented the strings of codes for the 20 amino acids.

In 1970, Hamilton O. Smith, working at the Johns Hopkins Medical School in Baltimore, isolated from cellular fluid for the first time a restriction enzyme, which is what cells produce when cutting DNA nucleotide sequences at specific locations to assist DNA repair, to inactivate invading viruses, or to conduct other cellular business. Smith received a Nobel Prize for this achievement.

In 1972, Paul Berg became the first scientist to construct a recombinant-DNA molecule—a molecule containing parts of DNA from different species. He combined a chromosome from a virus with genes from a bacterial chromosome and shared the Nobel Prize in chemistry in 1980 with the scientists who developed gene sequencing. In a brief autobiography written for the Nobel Prize committee, Berg noted, "My colleagues and I succeeded in developing a general way to join two DNAs together in vitro. . . . That work led to the emergence of

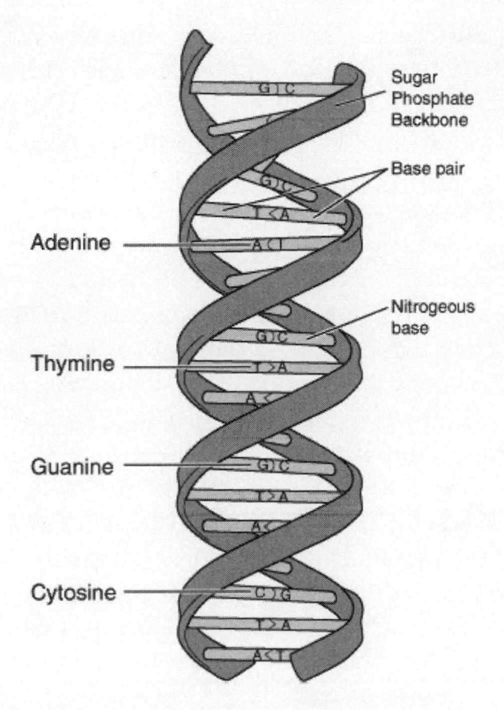

DNA strand. *Courtesy of the National Institutes of Health.*

the recombinant DNA technology thereby providing a major tool for analyzing mammalian gene structure and function and formed the basis for me receiving the 1980 Nobel Prize in Chemistry."

In 1973, Herbert Boyer of the University of California at San Francisco and Stanley Cohen of Stanford University learned simultaneously how to do genetic engineering based on "Ham" Smith's pioneering work, Berg's work, and the work of others. As Genome News Network summed it up nicely: "It was a creative synthesis of earlier research that made use of:

- Living organisms able to serve as carriers for genes from another organism.
- Enzymes to cleave and rejoin DNA fragments that contain such genes.
- DNA molecules from one organism precisely targeted and manipulated for insertion into the DNA of another organism."

Berg, Boyer, and Cohen—and Ham Smith—launched the biotechnology industry. Their work made it possible to take the nucleus of a cell containing an organism's DNA, manipulate the DNA, and generate copies of the genetically altered cell. Of course, it all had to be done by hand.

In 1977, Frederick Sanger of Cambridge University developed the method used most widely—until quite recently—to decode or sequence genes. According to Sanger in his brief autobiography for the Nobel committee, "Around 1960 I turned my attention to the nucleic acids, RNA and DNA. I developed methods for determining small sequences in RNA. The work culminated in the development of the 'dideoxy' technique for DNA sequencing around 1975. . . . The method has been improved and automated" for sequencing the human genome.

In the mid-1980s, the ability to make copies of strands of genetic code was revolutionized by PCR. A DNA polymerase is an enzyme produced naturally by cells for copying DNA sequences. PCR makes it possible to take a tiny fragment of genetic sequence from any source material and amplify it many millions of times to make the four letters of the genetic code easier to read. The technological revolution was akin to advancing from copying a single book of the Bible by hand to copying a thousand copies of the same book and having it nicely collated.

It was an incredible synthesis of existing ideas turned into something brand new and extremely powerful. What many people don't realize is that PCR was made practical by the discovery in 1968 of a heat-loving microbe from a hot spring at Yellowstone National Park. Ramunas Kondratas, curator at the Smithsonian's National Museum of American History, documented the discovery, development, commercialization, and applications of PCR technology. According to Kondratas from his summary:

> The Polymerase Chain Reaction (PCR) technique, invented in 1985 by Kary B. Mullis, allowed scientists to make millions of copies of a scarce sample of DNA. The technique has revolutionized many aspects of current research, including the diagnosis of genetic defects and the detection of the AIDS virus in human cells. The technique is also used by criminologists to link specific persons to samples of blood or hair via DNA comparison. PCR also affected evolutionary studies because large quantities of DNA can be manufactured from fossils containing but trace amounts.

Kary Mullis invented the PCR technique in 1985 while working as a chemist at the Cetus Corporation, a biotechnology firm in Emeryville, California. The procedure requires placing a small amount of the DNA containing the desired gene into a test tube. A large batch of loose nucleotides, which link into exact copies of the original gene, is also added to the tube. A pair of synthesized short DNA segments, that match segments on each side of the desired gene, is added. These "primers" find the right portion of the DNA, and serve as starting points for DNA copying. When the enzyme *Thermus aquaticus* (Taq) is added, the loose nucleotides lock into a DNA sequence dictated by the sequence of that target gene located between the two primers.

The test tube is heated, and the DNA's double helix separates into two strands. The DNA sequence of each strand of the helix is thus exposed and as the temperature is lowered the primers automatically bind to their complementary portions of the DNA sample.

Simultaneously, the enzyme links the loose nucleotides to the primer and to each of the separated DNA strands to incorporate the loose nucleotides in the appropriate sequence. One complete reaction takes about five minutes, yielding two double helixes containing the desired portion of the original. Each time the heating and cooling is repeated, the number of DNA copies is doubled. After 30 to 40 cycles are completed a single copy of a piece of DNA can be multiplied geometrically to hundreds of millions of copies.

Thermus aquaticus is a bacterial thermophile discovered by Thomas Brock and Hudson Freeze of Indiana University. Brock pioneered the study of heat-loving microbes throughout the 1960s and 1970s. But he assumed the microbes were an unusual type of bacteria that had adapted to the hot springs. Currently professor emeritus at the University of Wisconsin, Brock was the first microbiologist to devote his career to the study of microbes from hot springs. His living laboratory was Yellowstone National Park.

"At the time, I was interested in simple ecosystems where I could study microorganisms in their natural environments. In particular, the hot springs in Yellowstone seemed to be steady-state systems containing large and readily accessible microbial populations. These 'experiments in nature' soon led me to discover they are a habitat for bacteria that grow at high temperatures," Brock said. He discovered long-stranded, pink filamentous *Aquificales* and in 1965 he and his research partner isolated *Thermus aquaticus.* Brock's work at Yellowstone actually laid the foundation for several revolutions.

"In 1970–71, we described two new organisms from such hot, acidic environments, *Thermoplasma acidophilum* and *Sulfolobus acidocaldarius*. For some years, the main interest in these organisms was physiological and biochemical. However, in the late 1970s, Carl Woese of the University of Illinois reported that these two microorganisms are related to the methanogenic bacteria, thus greatly broadening his archaebacteria concept. Soon interest in thermoacidophiles broadened significantly with other researchers, particularly Wolfram Zillig and Karl Stetter in Germany, exploring other hot, acidic habitats for additional examples of archaebacteria," Brock said.

Brock published seminal research on hot springs microbes for 40 years and found the first living organisms colonizing piles of smoldering coal.

The Taq polymerase enzyme from *Thermus aquaticus* had been known about and studied since the 1970s. A significant improvement on Mullis's PCR invention came when people in his lab grew tired of the still tedious laboratory process involved initially with PCR. The original process involved a polymerase that broke down every time DNA was heated in one of the PCR cycles. This meant lab technicians had to add more polymerase by hand. The new idea was to apply the Taq polymerase of *Thermus aquaticus* to PCR. The organism thrives at temperatures of about 150 degrees F and its natural polymerase enzyme can withstand temperatures up 176 degrees F. Taq polymerase remained relatively stable as the DNA was repeatedly cooked, cooled, and copied, making it possible to automate PCR.

In the 1980s, Norman Pace saw value in PCR for the field of microbial ecology. Pace had been using the technique developed by Woese in the 1970s to identify microbes based on a segment of an RNA gene common to all or at least most microorganisms. This segment was from the 16s ribosomal RNA (16s rRNA) gene. Woese's method of hand-copying segments of ribosomal RNA from hoards of microorganisms was extremely tedious. After PCR became available, Pace's group combined it with Woese's 16s rRNA method of identifying microorganisms. Suddenly they possessed the ability to very rapidly identify 16s rRNA gene segments from raw DNA that was purified from a handful of dirt. They started cataloging species and comparing the genetic fingerprints of the RNA genes of many microorganisms. If they found a piece that had not been seen before, they probably had a new species. Pace bridged the two revolutions. However, while he could identify microbial species by 16s rRNA PCR, Pace could not determine the microbes' other genes with his methods.

In the late 1980s, Jay Short developed a new method of making copies of much larger segments of genes than had been possible. Short could take big pieces of DNA from a microorganism and place them in a cellular vehicle for copying genes, which scientists called a vector. The most commonly used vector was *E. coli* bacteria, which was amenable to accepting foreign DNA and then dividing into colonies that contained the new DNA. Short called this his "zapping method." Short's zapping method, combined with Leroy Hood's new sequencing machine developed in the 1980s, helped get the ball rolling for Venter to rapidly detect expressed human genes and soon afterward to sequence the whole genomes of microbes.

In the early 1990s, Venter synthesized these technologies to create the means to find pieces of active human genes in tissue samples, which could be sequenced relatively rapidly, using PCR and the automated sequencing machine. The gene fragments were called "expressed sequence tags" (ESTs). This was revolutionary, but it made Venter rather unpopular with people running the Human Genome Project. They had their own way of doing things and ignored, even cruelly ridiculed, Venter's ESTs. After founding The Institute for Genomic Research (TIGR) in 1992, Venter began working on methods of applying his ESTs to decode the entire genomes of individual microorganisms. He did this by refining a method known as "shotgun sequencing," which had been developed years earlier for single gene sequencing. In the early 1980s people could spend two years sequencing only 500 base pairs of DNA. Shotgun sequencing was an approach that obviated the need to laboriously lay out each segment of DNA in preparation for sequencing. Shotgun sequencing selects DNA segments at random for sequencing and worries about ordering later. This simple approach has the beauty of providing data early, but has the unfortunate drawback of requiring that each segment of DNA sequence be pieced together like a jigsaw puzzle to determine the full gene sequence. To further complicate matters, the random selection of segments for sequencing meant that a lot of DNA had to be sequenced again toward the end of a project. In fact, more than 10 times the amount of sequencing was often required to fully sequence a gene.

All in all, shotgun sequencing worked at the gene level, but few thought this approach was practical for sequencing an entire genome, let alone a massive human genome. However, technology was moving fast. Sequencing vectors, which were used to copy segments of DNA, were rapidly evolving. Short created a vector, which he called the "transforming lambda ZAP system." He developed the system while working at Stratagene. The lambda ZAP vector was used by Venter for early EST sequencing. Combined with improving PCR technologies and automated sequencing, Venter thought the speed of these new systems would allow shotgun sequencing to conquer a genome and possibly the human genome. The microbial genome became a wonderful proving ground and provided a path for funding. His strategy was criticized broadly in view of the more certain ordered approach, but the bet fit his personality and may have been reasonable since he didn't have another way of getting ahead of the established gene mapping in-crowd.

While Venter began applying shotgun sequencing to microbial genomes, microbial ecologists were rapidly progressing with 16s rRNA and PCR to genetically fingerprint microbes in environmental samples. This was revealing a vast hidden universe of microbes, but to study individual microbes they still had to be isolated and cultured.

Short redesigned the whole microbial game in 1994 by proving that the genes of entire populations could be cloned (copied) and expressed in his zapping vectors, which was not possible by PCR. Pace's method could provide information for identifying microbes and constructing family trees. But to learn more about microorganisms in the environment, an approach was needed to stably clone genes directly from raw DNA and see if the exotic genes could be expressed without killing the domesticated host cell required for producing clones. Short knew lambda was a good candidate for this challenge. It had been proven to be an efficient cloning system. His team, however, did not believe it would work on raw DNA from the environment. They were surprised when it worked even better than Short expected. His method became the first to allow harvesting and expression screening of DNA directly from the environment. The direct cloning and stable expression of genes to explore previously unknown microorganisms from other domains and extremes of life opened up a brand new world. It allowed the identification of millions of novel genes, and not in one microbe at a time, but thousands at a time.

"There are more genes in a handful of soil than in the entire human genome," Short said just before the Costa Rica trip. "We know how to find them, tell you what they do, and enhance them. All we need to start with is raw, uncultured microbial DNA."

This was an important approach at the time because the challenge of sequencing environmental communities containing millions of species of microbes was about a thousand times greater than that of sequencing the human genome and a million times greater than conquering a single microbial genome. Applying straight sequencing without expression screening during that time would have had limited success since it took six months to a year to sequence a single microbe. Short wanted to screen the genes of a million different species of microbes at once to find the best genes that nature could provide for specific industrial tasks.

The only problem was that Short's method was something so completely different that few scientists understood what Short had achieved.

"A few people were harvesting DNA from the environment but focused on screening for known genes, primarily 16s rRNA genes. Nobody had dreamed that you could take genes from archaea, get all of those archaea genes to switch on and become expressed in a bacterial cloning system of the day. Many on my own team said it would never work. But as head of research I could insist that we try and we did it and it worked, a little better than I expected," Short said.

Short applied his lambda ZAP vector system to the environment at Diversa. It is able to express a billion genes per day from the raw DNA of microbes. By comparison, the entire human genome has only about 30,000 genes. As sequencing speeds advanced, Short initiated the first direct sequencing of the same environmental libraries, skipping the expression, PCR, and other selections. The gene-based study of microbial communities, whether through the 16s rRNA PCR approaches of Pace or the gene- and genome-based approaches of Short, would later be recast as the new field of metagenomics.

Short and Mathur have a good history of invention. In 1991, the duo was working at Stratagene and had become frustrated with the Taq polymerase that revolutionized PCR. Short discussed with Mathur how they needed a polymerase that was even more stable and able to work at higher temperatures than Taq polymerase. Six months later Mathur came across a paper published in 1986 in the *Archives of Microbiology* about an archaea named *Pyrococcus furiosus*—fireball. The microbe thrives at temperatures of 212 degrees F. Karl Stetter discovered it in volcanically heated marine sediments on a beach at Puerto Levante on Vulcano Island in Sicily—Stetter's favorite hunting grounds and family vacation spot. Mathur wondered whether the DNA polymerase of Stetter's bug would further improve PCR. Mathur began experiments and found that *Pyrococcus* was much more thermally stable than *Thermus aquaticus* and resulted in fewer errors in copying DNA sequences. "I have really only made one big scientific discovery so far in my life, and that was the Pfu DNA polymerase proofreading thermostable enzyme used for PCR. It is still Stratagene's biggest product with about $20 million in annual sales. I'm very proud of this accomplishment," Mathur said. The company patented the *Pyrococcus* DNA polymerase, and they published a paper on it in 1991. Stratagene began getting its own share of PCR-related royalties.

Stetter told me he was furious about *P. furiosus* being stolen from under his nose "by this fellow Eric Mathur." The two scientists would not meet until 1994 when Stetter and his collection of cultured archaea were brought in by venture capitalist Wallace Steinberg to co-found Industrial Genome Sciences/Diversa. At the beginning when Mathur was invited to join the company, Stetter said he felt it would be best to have the nemesis who outsmarted him under the same roof.

Diversa's goal was to find or create novel enzymes that could be genetically engineered, grown in vast quantities, and then sold to industry partners. That was simple in concept. Enzymes are the workhorses of biological reactions—the catalysts critical to industrial processes such as converting corn into ethanol and the manufacturing of pharmaceuticals. The archaea and bacteria from extreme environments operate in environments very similar to those that take place when chemicals are being manufactured and the conditions during many industrial processes. The microbial industrial complex does naturally what the human industrial complex would like to do more cheaply and efficiently. Short said that industry will stop polluting the environment when it costs them less money to do so. Short says microbe-based enzymes are one way to lower costs and reduce pollutants produced by industry. Sounds reasonable.

During the 1990s, Diversa had developed several key partners in industry, government, and academia based on Short's methods of accessing nearly 100% of microbial genes in environmental samples and evolving their DNA. Short was the first to develop and apply high throughput expression of genes and sequencing to microbial genome populations. This was far different from the classical characterization work of 16s rRNA sequencing and shotgun sequencing. The new method became very attractive to industry. The company's industry collaborators include Novartis; BASF; Cargill Health and Food Technologies; DuPont Bio-Based Materials; Medarex, Inc.; Merck & Co., Inc.; Xoma, Ltd.; Bayer Animal Health; DSM Pharma Chemicals; The Dow Chemical Company; and Givaudan. Diversa currently has six products on the market that are expected to reach more than $80 million in sales in 2008.

Diversa also had a great friend and partner at the DOE, which officially got into the bug hunt in 1994. The DOE immediately began collaborating with TIGR and soon afterward with Diversa. The DOE launched the Microbial Genome Initiative in September 1994 as a companion to the Human Genome Project. By the time of the Costa Rica expedition, Diversa and the DOE were preparing to announce a major collaboration to advance the discovery of microbial genes in environmental samples. They were calling it the environmental equivalent of the Human Genome Project.

The DOE was funding TIGR to sequence the whole genomes of microorganisms it believed would be important to industry and for cleaning up toxic and radioactive waste left behind from the manufacturing of atomic weapons. The DOE went straight away to the outer limits to secure the genomes of two of the most extreme archaea and a bacterium that were colonizing toxic waste sites. In the microbial genome initiative's first year, it funded the sequencing of

- *Pyrococcus furiosus*, the aforementioned hyperthermophile that thrives at the boiling point of water.
- *Methanococcus jannaschii*, the headline grabber in 1996.
- *Methanobacterium thermoautotrophicum*, a sewage sludge methanogen that grows optimally at 65 degrees C. It uses carbon dioxide to break down sludge and gives off hydrogen and methane.

In 2001, the DOE launched Genomes to Life (GTL) to further learn how microbes function and how they can be exploited for energy, medicine, and industry. GTL also encouraged the budding field of comparative genomics. Comparisons of the whole genomes of different organisms reveal a lot of basic information about gene families, natural selection, and origins.

It is a grand thing to sequence an entire organism. But it is an entirely different game to comprehend a dynamic ecosystem of microorganisms containing hundreds or thousands of interacting microbial species. Shotgun sequencing of a single microbe is kind of like turning the Hubble telescope toward Earth and taking hundreds of millions of pictures of three-meter squares of the entire surface and then piecing them together based on similarities between the edges of the pieces.

Short's team was redesigning these technologies to forge a new path for simultaneously conquering millions of hidden microbial genomes. They did this by taking tiny slivers of raw material from each one of those three-meter squares. If they wanted an enzyme for a pulp and paper plant they would find a pH 10 environment with wood sticking out of it. They would harvest DNA from it and then make environmental libraries and identify genes that encode proteins that align with the industrial conditions for the targeted application. Always hoping for surprises, Short required that they salt in other random environmental samples. What was the difference here? Being able to look at millions of different organisms and their genes moved everything into a new world. These genes all make proteins of some variety. After sorting all the families of related genes, independent of their species, into groups, Short developed technology to rapidly rearrange the genetic sequences that encoded all of the recovered proteins to create something nature had not yet created.

"Rather than looking at the world one piece at time, as with sequencing, we could look at the whole world at once and pick out the pieces we want," Short said.

Short's methods provided essential metabolic information about microbes missing from sequencing alone. On September 26, 2002, Diversa announced that it had been awarded a patent on its method of amplifying and sequencing the DNA of nearly 100% of the organisms in an environmental sample. In a single year, Diversa was awarded 20 patents on its new technologies for identifying and evolving the genes of microbes. By the time I left for Costa Rica, Diversa's patented method had been used to develop gene libraries of tens of thousands of novel genes from previously unidentified microorganisms.

Microbial genes and their proteins are at the heart of the future of biotechnology. Microbes naturally make most of the coolest and weirdest enzymes known to science. Most enzymes are made of proteins. Proteins are made of amino acids, also called "peptides." The 20 amino acids form different sequences similar to the way the four basic nucleotides of genetic code form sequences. With 20 different amino acids, a lot of different permutations can arise. Different combinations of amino acids form every type of protein known to science. It is the sequence of the amino acids that determines the protein's function. Changing a single peptide in the amino acid sequence can dramatically alter the function of a protein.

After identifying 100% of the species in an environmental sample, the next step is the "directed evolution" technique in which the amino acid sequences of the best of nature's proteins are recombined to see whether a protein's natural function could be enhanced or, even more remarkably, be made to do something entirely new to nature. In addition, typically every single amino acid would be mutated at every position in the protein to look for further improvements. This process alone involves thousands of mutations and would have required about 100 scientists each working a full year to complete less than a decade earlier. Now it can be accomplished in less than two weeks.

Short says the company's strategy is analogous to what Kentucky breeders do to get the ultimate racing horse. "They find the best animals possible and breed them to come up with an even greater animal. They don't start out with two mules. We start out with the best microbial genes that nature has to offer and then do our own magic," he says.

The gene libraries allow Short to select specific metabolic traits and combine them into a designer enzyme for a corporate partner's needs. By 2003, Diversa had come up with several real evolved genes. They found on one gene a string of sequences that offers very high temperature resistance. They found on another gene a string that codes for an enzyme that breaks down cornstarch with high productivity. Normal enzymes that do this don't work well at the higher temperatures needed to efficiently break the starch down. Put the two gene segments together and a gene is created that confers high temperature resistance and built-in enzymes for breaking down cornstarch. This saves money and time for people making bioethanol fuel. In 2003, Diversa collaborated with Syngenta to develop a heat-tolerant digestive enzyme that Syngenta would genetically engineer into seed to grow corn that would express high levels of the enzyme, which Short calls "Amylase." The idea is that Amylase-enhanced corn will make it simpler and less expensive to convert cornstarch into ethanol by eliminating the need to add liquid enzymes during the conversion.

In the late 1990s Diversa began manufacturing a digestive enzyme made from bacterial genes that is added to livestock feeds to aid digestion in pigs and chickens. Phosphorus exists naturally in commercial poultry and pig feeds. But pigs and chickens absorb only about 30% of the natural phosphorus in feed so inorganic phosphorus is added to enhance nutrition. Adding inorganic phosphorus to feed is expensive and is one of the most prolific pollutants of streams, rivers, estuaries, and bays. Much of the inorganic phosphorus added to feed remains undigested and is excreted as waste. The phosphorus-rich waste then runs off into water systems, such as the Chesapeake Bay, which is surrounded by poultry farms. The run-off results in algae blooms, displaces oxygen levels in fresh water, and fouls drinking water supplies.

Diversa developed a series of products called Quantum and Phyzyme XP to improve the digestibility of phosphorus and other nutrients naturally contained in animal feed. According to Short, the benefit is that livestock can achieve higher rates of nutrition, allowing farmers to reduce the amount of phosphates that are added to feed to aid digestion. Lower phosphate levels translate into less pollution of streams near commercial farms.

Among the other products resulting from Diversa's many expeditions is a new naturally derived ice cream flavor for Givaudan based on microbes found in a rotting coconut in the jungles of Ghana.

Of particular interest to me was that Diversa had developed a biodiversity agreement with the National Park Service in 1997 through John D. Varley, director of scientific research at Yellowstone National Park.[2] The agreement became the model used by Yellowstone and all companies and academic teams working in the park's hot springs. Yellowstone has long been a focus of bioprospecting. Microbiologists have been curious about the hot springs since the 1920s. With Varley, Diversa had taken a bold step into the Wild West free-for-all of bioprospecting. Short insisted that royalty-sharing contracts be worked out ahead of any bioprospecting trip. What made this bold is Short was seeking to pay the royalties rather than receive them. The Yellowstone agreement was considered important enough that Vice President Al Gore attended the ceremony and announced it at the park's 125th-year anniversary in August 1997.

The Diversa/Yellowstone agreement was a milestone for the National Park system. Leif Christoffersen had been working on the bioprospecting agreement since 1995 before coming to Diversa in 2000.

In 1995, Christoffersen was a program manager for the World Foundation for Environment and Development (WFED). He began working on a bioprospecting program at the WFED for the National Park Service with Yellowstone's John Varley. Christoffersen and Preston Scott, executive director at the WFED, laid all of the groundwork that led to the ultimate agreement signed between Diversa and Yellowstone National Park. Private companies, especially pharmaceutical giants, had been bioprospecting around the world for decades, taking samples back to their labs and developing new drugs without making royalty-sharing arrangements with the owners of the resources. This is called "biopiracy." After applying *Thermus aquaticus* as the keystone of PCR, Cetus sold the technology to Roche Molecular Systems for $300 million. Had even a small royalty agreement been in place when PCR was developed, Yellowstone would have become more economically self-sufficient and could have dramatically enhanced its own scientific research and park services, according to Varley. Yellowstone National Park never received a dime in royalties.

"The Yellowstone/Diversa cooperative research and development agreement, in distinguishing between the 'resource use' and 'research results,' helps to assure that Yellowstone shares in profits and scientific information generated from valuable research results," Varley said. "The agreement grants no exclusive or private rights with respect to any biological specimens collected from Yellowstone."

This prevents a single company, including Diversa, from developing a monopoly on any of the park's resources. A company cannot own any organism; it can only own the improvements to the genes or to the natural DNA polymerase enzyme on which PCR was based. The Yellowstone/Diversa agreement also ensures that bioprospectors share their scientific data on whatever they discover about microbes sampled from anywhere in the park. This is significant because private companies are accustomed to keeping their data secret.

After Short was promoted to CEO of Diversa in 1998, he aggressively pursued the biodiversity agreement with Yellowstone as a model for all of the company's business. Short said Diversa needed to establish a precedent for bioprospecting for two reasons: Ethical bioprospecting is good for business. But exploiting the biological resources of countries and indigenous people for profits is not morally right and obviously bad for a company's image.

Pragmatically, one of the last things a company wants to have happen is develop a cure for cancer or a new biofuel from a termite microbe found in Costa Rica, or any nation, state, indigenous region, city, park, or backyard, and then be sued the moment the company starts turning a profit. If you are beholden to Wall Street, the stock will tank as soon as the plaintiff's lawyers start filing briefs. Diversa's effort to nail down a single effective biodiversity agreement stemmed from an international treaty adopted at the Earth Summit in June 1992 in Rio de Janeiro.[3] This Convention on Biodiversity demanded accountability from companies, protection and sustainable use of the natural resources that were being collected, and some type of profit or royalty-sharing arrangement with the countries. The United Nations has strengthened the treaty since 1992 and most countries, except for the United States, have ratified the international agreement. The Diversa guys were savvy scientists and businessmen but I also had the sense they were honest and actually cared about this stuff, especially since few other companies were taking this path.

I arrived at the Hotel Bougainvillea in San Jose in the late afternoon and finished reading the background materials while waiting for the Diversa team to arrive. At 10:30 p.m. the phone in my room rang. The group was in the lobby for a brief meeting. Mathur was grinning ear to ear, nearly sparking with electricity. Briggs was a bit reserved, soaking it in. Christoffersen was polite, friendly, can-do. Hillary Theakston, Diversa's public relations person, seemed poised to ensure that Mathur did not say anything that wasn't corporate-approved. She had to give up as soon as Mathur ordered the first round of beers. At 11:15, Theakston stood up and suggested everyone get to bed since we had an early start. No one moved. Mathur wished her goodnight. As soon as the elevator door closed behind her, Briggs suggested we have a look at the casino in the hotel. Mathur talked rapid-fire about termites and biofuels and expeditions as we played blackjack and flirted with the dealers. At 2 a.m. we went to bed to catch about four hours' sleep. It was going to be an interesting journey.

INBio is located at the southern outskirts of San Jose. It's a tourist destination popular with Costa Ricans. The facility has a large park with trails that wind through different collections of plants, trees, and vegetation from the rainforest, the Central Valley, the dry forest, and wetlands. After touring the labs and the park, our group and four INBio scientists piled our gear into two aging Land Rovers and headed east for Limon on the Caribbean coast and on southward to our jungle resort lodgings near Cahuita. After a night of drinking Cuba libres and chatting "til all hours," it was nearly 9 a.m. the next morning before the group departed for a nearby section of jungle trails that abutted the Caribbean coast. The sun was blistering hot when we parked the Land Rovers, unloaded our gear, and marched single file into the humid, coastal rainforest. Moss-covered kapok trees graciously blocked the mid-morning sun. Dense rows of raffia and coconut palms lined both sides of the trail, which we shared with thousands of very industrious leaf-cutter ants.

Miriam Hernandez, INBio's head of molecular biology and bioprospecting, stopped ahead at a large, round, termite nest attached to a dead coconut palm to begin taking samples. (See Plate 2.) Hernandez removed a small shovel from her backpack and dug into the side of the nest. Dozens of termites swarm onto her hand. She grinned as the insects crawled over her skin, then she scooped some of the termites into a jar to add to her collection of beetles, scorpions, moths, and spiders back at INBio. Mathur had been burning for months with the desire to extract microbes from the hindguts of termites. Termites inhabit nearly 70% of Earth's surface. They are one of the planet's most efficient bioreactors, capable in one day of producing two liters of hydrogen from fermenting a single sheet of paper. Our scientists can split atoms and develop all types of ways to destroy life, but they cannot figure out how to make hydrogen as efficiently as a termite. Therefore termites are high-priority bug items for the DOE and a number of biotechnology companies. As a bioreactor, the little termite is one of the most significant natural contributors of greenhouse gases in the atmosphere. It's their job and Earth can handle it—all part of the current natural balance of the carbon cycle. In nature's industry, termites are the masters at degrading old trees and forest litter. But termites do not work alone. They are actually the heavy machinery, the vehicles, which carry a complex of microorganisms that do the real work.

Jared Leadbetter has done most of the basic research on termite hindgut microbes and was Mathur's inspiration for Diversa's termite project. Leadbetter, an assistant professor at the California Institute of Technology, said in a radio interview with Jim Metzner during a show called *Pulse of the Planet*, "Many people, when they first observe the termite hindgut community, are really shocked. It's like being an ecotourist in *Alice in Wonderland*. It is so wild a place. There are these objects that look like seals: hairy, long, furry animals, which are swimming amongst all these other cells. Of course, these are single-celled microbes. But they have such bizarre shapes and movement in their liquid environment that they invoke much more complex, larger animals in our mind. So, it's just a very striking sight when you look at these organisms that have been freshly collected from the termite hindgut."

Using Diversa's environmental libraries, the Joint Genome Institute (JGI) was to sequence the entire community of hindgut microbes to determine the biochemical pathways involved in converting cellulose to energy. Mathur says it is not far-fetched to consider that biochemical pathways in the guts of termites could in the near future lead to biological production of hydrogen energy and ethanol. Producing the hydrogen biologically actually is no problem. It's figuring out how to store it. But one day perhaps, instead of petroleum refineries, immense vats of microbes will line the New Jersey Turnpike.

As the team walks deeper into the jungle, Mathur tells the group to keep their eyes out for color gradients in the soil because these are likely to be caused by colonies of microbes. He also says to be alert for pools of foul-smelling water that have collected on the floor of the rainforest and especially inside the wells of bromeliads. He wants to sample everything. Why? I ask Mathur to explain again what Diversa does for a living.

"The samples we collect typically consist of less than 50 grams of soil, sediment, leaf litter, or other environmental materials. The total microbial-community nucleic acids are first extracted from the samples and then converted into highly complex, representative metagenome libraries," Mathur said.

A metagenome library is the library of all of the genes found in a particular environmental sample or collection of samples. Depending on the discovery application, the environmental libraries are sized to contain gene fragments ranging from 3,000 to 150,000 base pairs. These pairs are the nucleotides that bind to each other from opposite sides of the strands of DNA. Base pairs form the rungs of the ladder of the double helix.

"The libraries must have the complexity to capture the majority of genes from the genomes of more than 15,000 unique microorganisms. Diversa currently possesses more than 4,000 of these metagenome libraries. That is a vast resource and repository of genetic material for enzyme discovery. Since the libraries can be propagated within our labs we rarely need to go back to the environment to resample, which minimizes our impact on the environment," Mathur said.

To identify and recover enzyme candidates that fit a particular industrial performance profile, Diversa uses computers to sift through the metagenome libraries to find the genes and gene products of interest. The term "applied metagenomics" refers to combining gene expression, culturing and applying sequencing technologies to raw DNA from uncultured microorganisms.

"Shotgun sequencing of metagenome libraries as a method of gene discovery has become more practical, as the cost of sequencing at genomic centers, particularly JGI, is now less than one-hundredth of a cent per base. Can you believe that? That is ridiculously cheap, man. But the real art lies in how we recover the proteins of genes," Mathur said.

Screening one of Diversa's metagenome libraries using traditional agar culture methods would require 10,000 petri plates, each containing 10,000 genes or gene fragments.

"This just isn't practical in an industrial setting. One example of how we solved the throughput challenge is the development of an ultrahigh throughput screening platform called 'Gigamatrix,'" Mathur said.

The Gigamatrix system uses a petri-dish type of platform that contains 400,000 tiny wells in which genes are expressed. Screening one metagenome library of 10,000 to 15,000 genes with the Gigamatrix can be done in less than three hours.

"Another ultrahigh throughput proprietary screening platform is what we call 'SingleCell.' It is based on fluorescence-activated cell sorting. Diversa has developed and optimized the technology for sorting and screening recombinant bacterial cells, using either functional or sequence-based screens, at rates exceeding 10^7 cells per second," Mathur said.

I stopped taking notes and started using the tape recorder in my daypack. I stared anxiously at the recorder to make sure it was working. I rarely used one as a reporter. I don't trust them.

"Am I talking too fast, man?" Mathur grinned. Briggs gave me a sympathetic look. "He's like this all the time," Briggs said.

We found a spot to sit where no ants would crawl into our pants. I related my phone conversation with Short before the trip and asked Mathur to explain the meaning of "Gene Site Saturation Mutagenesis" and "Tunable Gene Reassembly."

"Okay. This is what Jay designed to optimize our enzyme candidates. Gene Site Saturation Mutagenesis creates a family of related genes that all differ from the parent gene by at least a single amino acid change at any defined position. This lets us produce all possible amino acid substitutions at every position within a chain of peptides. The family of variant genes created using this technology is then available to be screened for enzymes or even antibodies with improved qualities, such as the increased ability to work at high temperatures or resistance to certain chemicals, or other properties important in specific industrial processes. Beneficial mutations can then be combined to create a single, highly improved version of the protein. Tunable gene reassembly allows blending of gene sequences. You can actually create genes that don't exist in nature that are optimized for specific industrial processes," he said.

At the end of the day the sweaty, dirt-smudged team packed the sampling equipment back into the Land Rovers to return to the lodge. But about halfway home we made a detour onto a small dirt road that ended at the edge of another jungle trail. Hernandez and Christoffersen removed a white sheet from one of their boxes and stretched it onto a frame. They placed battery-powered spotlights behind it and turned them on. (See Plate 3.)

"Bugs." Mathur said. "They're attracted to the light and hang onto the sheet. We'll come back after dinner and collect them. I want at least one of everything that lives here. Most of these insects have microbes as symbionts just like the termites."

Back at the lodge, everyone showers and finds clean clothes. After a fresh seafood dinner in Cahuita we return to the trail to grab whatever insects have fallen for the light-behind-the-sheet trick and scout for more along the dark trail. After a bit, Christoffersen and Hernandez return, quite pleased. She holds up a small glass vial and shines her flashlight on it. Inside is a tiny black scorpion. She says it's the most poisonous species in Costa Rica even though it's the smallest.

"Can I have that?" Mathur asks. "I'll give it back, but I'd like to keep it for awhile."

Mathur is like a kid. He started taunting people with the scorpion. "I feel really powerful," he said. He put the vial in his shirt pocket. Later in the evening, when the rum was gone and people were heading back to their beds, Mathur asked, "Are there any bars around here? Leif, go ask the Costa Ricans if there are bars within 10 miles or so."

Our sleepy driver came to the cottage where Mathur was again looking at the scorpion. Christoffersen announced there was a bar about five miles away on the beach in Cahuita. The driver took us. Five miles on a small jungle road took some time. Along the way, Mathur kept asking Christoffersen, "Is he sure there's a bar out here?"

Christoffersen replied, "He's not sure it's open but he knows the bar is there."

"How do you say in Spanish, 'If you fail, you will die'?" Mathur asked.

"Si falles, mueres," Christoffersen said.

The driver looked back nervously at Mathur, who removed the vial with the scorpion from his pocket and held it to the driver's neck. "Si falles, mueres!" Mathur howled with laughter. The driver glanced at Christoffersen, who gave him a reassuring look.

Spending time with people under rugged conditions tends to bring out their true character. Mathur is candid, gregarious, and absolutely fun. As we made our way through the jungle the next morning he took off his pith helmet, revealing a freshly shaved head covered by a bandana. I can't say I have ever seen anyone actually wear a pith helmet outside of a Tarzan movie, but it looked natural on him.

"Bacteria are the most abundant and creative form of life on Earth. They can evolve to live almost anywhere," Mathur said, as we began climbing inside a muddy cave. Briggs added that microbes have had the opportunity to evolve their genes for at least 3.5 billion years, working out a tremendous diversity of biological processes. Now the tools exist to exploit these natural biochemical pathways and to adapt the metabolisms of microbes to all kinds of industrial processes. Having the ability to discover vast amounts of new microbial species from environmental samples has opened an entirely new universe.

"I have absolutely no doubt that microbial genomics will be the great revolution of the 21st century in the same way physics was for the 20th century," Briggs said.

On the last afternoon of the expedition Mathur suggested we go to the beach. The sky was a crystal blue, and the Caribbean was just cool enough to keep us from roasting under the sun. Briggs was as pale as a nude laboratory mouse. As we cooled off in the sea Mathur soaked his bandana and wrapped it around his head to keep it from burning. He could be a pirate. Briggs's pale shoulders turned pink, and he covered himself with his white T-shirt. Mathur began talking about how much everyone in microbiology owes to Carl Woese's discovery of archaea. Briggs and Mathur discussed how quickly the discovery of archaea stimulated the development of new science and technology. Briggs said the realization that life thrives in extreme environments revolutionized microbiology and biotechnology. It was also rapidly altering concepts about the origin of life. Briggs said he was amazed more people weren't aware of what was occurring.

I tingled with excitement. Mathur and Briggs could see I was hooked. Reporters love untold stories, especially ones that involve brilliant, nice guys who get screwed by the big important scientists of the day, but win in the end. Briggs explained how Woese's work was ignored and ridiculed for at least a decade, except by a handful of scientists that included Diversa's co-founder Karl Stetter. Mathur chimed in, adding if it hadn't been for Woese, scientists would not have uncloaked the invisible universe. There would be no Diversa. No Genomes to Life. No real excitement about the possibility of life beyond Earth. Mathur became more animated as he spoke. Then he paused, looked up at the sky, and stated with utter confidence:

"One of the most profound scientific breakthroughs of the 20th century, maybe in all of history next to Galileo, has gone almost completely unrecognized by the public."

I glanced cautiously at Briggs. He nodded in agreement.

"We're talking about changing the way people view the origin of life on Earth—that extreme species of archaea and bacteria could very well survive in extraterrestrial environments and appear to be directly related to the first forms of life on Earth and are still here evolving in incredible ways. We're talking about harnessing these organisms to revamp the way we make everything from energy to pharmaceuticals to household products and industrial chemicals. I'm dead serious, man. Hardly anybody knows about this."

Briggs was still nodding. "I hadn't thought about it, but it's true."

Mathur looked at me and declared, "You have to call Carl Woese and talk to him. You have to meet Jay and Karl Stetter, all these people, and come with us when we go to Kamchatka in Siberia and to Yellowstone. Come bug hunting, man."

"You really should," Briggs said. "These are pretty remarkable people."

Yes. I was dumbstruck. How had I missed this story all these years?

I had to call Carl Woese.

3
WHAT THE HECK IS IT?

Carl Woese discovers a new branch of life and scientists try to make him suffer.

All wisdom is rooted in learning to call things by the right name. When things are properly identified, they fall into natural categories and understanding becomes orderly.

Confucius

Every story must have a beginning; every revolution a leader. So it was in the spring of 1975 when a 47-year-old molecular biologist named Carl Woese of the University of Illinois was handed some microbes isolated from sewage sludge, thus beginning what may be one of the most important revolutions in biology since Charles Darwin published *The Origin of Species* on November 24, 1859.

Carl Woese discovered that these microbes represented a new form of life—archaea—and brought us closer than anyone in history to uncovering the roots of the tree of life. He single-handedly opened for exploration the vast microbial universe to a new breed of scientists armed with powerful genomic technologies, forever changing our understanding of the diversity of life on Earth. With bare bones equipment and nothing automated in his lab except for an old turntable for jazz records, he invented the methods adapted today to discover untold numbers of microbial species in every conceivable environment. He opened our minds to the possibility—I would venture probability—of life beyond our planet. And when the biotechnology industry eventually develops the means to harness biofuels and manipulate microbes for all manner of industrial and pharmaceutical processes, it will trace the development to the door of Carl Woese's original lab covered with antiquated light boxes on the third floor of a red brick building on the campus of the University of Illinois.

Yet, for all of Woese's accomplishments, sludge would foreshadow how U.S. scientists would treat him for nearly two decades for uprooting the tree of life. One does not mess with the tree. But like Thomas Gold, whose Deep Hot Biosphere was inspired in part by Woese's pioneering work, Woese was not afraid to question dogma whenever he stepped in it and realized something did not smell right.

For nearly half a century, the ancient trunk of this stalwart tree has divided life into two clearly distinct branches representing prokaryotes and eukaryotes. Prokaryotes represented a super-kingdom consisting of all microbes that do not possess a cell nucleus. The second super-kingdom, which biologists believe branched off from the root somewhat later than the first, bears the much more colorful and charismatic eukaryotes. All eukaryotes have cells with a nucleus and comprise every other form of life from the little green globes of algae to ferns, flowers, and palmettos; cats, dogs, and dolphins; and ultimately the beautiful human being you see every morning in the mirror. The root of the tree represented a mysterious and ancient common ancestor of the prokaryote and eukaryote branches. And that was it. One is either missing a nucleus and is a prokaryote or one possesses a nucleus and is a eukaryote. It did not get much simpler than this, or so everyone thought.

Microbiology texts claim that this prokaryote–eukaryote division was proposed in 1937 by Edouard Chatton. This does not appear to be true. Chatton merely scribbled the term "prokaryote" on a single diagram unearthed quite recently during some detective work conducted by Woese.[1] Chatton never proposed prokaryotes as a new phylogenetic class, Woese said. Nevertheless, by the late 1960s when Woese began attempting to classify microorganisms based on his unusual new system of examining specific coding segments of microbial RNA, the prokaryote–eukaryote dichotomy had become the status quo of microbiology. The prokaryote–eukaryote concept is still the worldview of most microbiologists, much to the frustration of Woese and his closest disciple Norman Pace.

Classification of life is a quite messy, often nasty, and generally confusing business. So, let's make it clear right away that Woese and

Pace currently divide life into three domains consisting of bacteria, eukaryotes, and archaea. They wish to rid microbiology entirely of the term prokaryote, which they say was created as a matter of convenience to cover up gaping holes in the knowledge of how to classify microbes. When Woese first proposed that archaea represented a new form of life in 1977,[2] he referred to them as a new super-kingdom, called archaebacteria. By 1990, Woese adopted the term "domain" for the three new branches of the tree of life and shortened the name archaebacteria to archaea. His original term, archaebacteria, obviously made people think the new organism was a type of bacteria, which it was not.[3] (See Plate 3.)

Classifying species had always been done generally based on how a plant or animal looked and behaved, and whether or not two individuals could breed. This simply did not work for microbes, so by the 1960s microbiologists had seized upon the simple idea of classifying them based on whether or not they had a cell nucleus, which was the only apparent major difference between microbes as they appeared under a microscope.

Charles Robertson at Pace's lab at the University of Colorado quite recently used supercomputers and a sophisticated algorithm to analyze the genetic sequences of all known microbes. Everything falls into the domains of bacteria, eukaryotes, or archaea. Perhaps a fourth or fifth domain will be discovered by others in the future, but for now we've got three solid branches on the tree of life and no use for the concept of a prokaryote. Prokaryotes are a fabrication. They do not exist.

The two–super-kingdom classification of prokaryotes and eukaryotes introduced in the 20th century was a tidy advance over Ernst Haeckel's three-kingdom system of Plantae, Animalia, and Protista proffered to science in 1866. In Haeckel's day, kingdoms were the pinnacle. Haeckel's system represented an evolution in appreciation of the curious animalcules—microorganisms—by grouping them into a single kingdom. Before Haeckel's 19th-century paradigm, the scientific world relied upon the classification system developed by Carolus Linneaus who in his *Systemae Naturae*, published in 1735, divided everything in the world into three kingdoms—Animalia, Vegetabilia, and Mineralia—and into five ranks—class, order, genus, species, and variety. Microorganisms were organized into the Animalia and Vegetabilia kingdoms based on their motility, so the protozoans that darted around "prettily," as Leeuwenhoek wrote in 1674, beneath the lenses of his microscopes were Animalia; green globes from the lake and the curious bacteria found in scrapings of dental plaque from his wife, daughter, and a few old men were classified as Vegetabilia.

When Carl Woese proposed his third super-kingdom in 1977, the scientific tree of life had been altered only twice since the Age of Enlightenment. Woese had no intention of messing with the tree. Woese had accepted the two–super-kingdom system as long as it made sense. Neither did he set out to find a new and bizarre form of life. Woese was after much bigger game: the origin of life. To truly understand origins, one must possess a complete tree of life and no such tree existed for microorganisms. As Woese began probing the evolutionary path of microbes, he found that the prokaryote–eukaryote phylogeny was essentially meaningless. Leaders in microbiology during Woese's formative years as a scientist claimed that a molecular tree of prokaryotes would be impossible to construct. Woese says microbiology as a discipline gave up trying to determine the bigger evolutionary picture.[4]

In the early 1960s, Woese became obsessed with categorizing microorganisms based on their molecular structure. He quite insightfully believed that RNA, which scientists were just beginning to understand, contained fundamental secrets to the origin of cells—and thus of life. Woese said he was determined "to move the evolutionary discussion away from animals and plants and onto the molecular level, where it belonged in the 20th century." Woese insisted that without a molecular taxonomy of microorganisms, understanding the true nature of evolution and the origins of life would be impossible. Microorganisms were generally thought to be the ancestors of life on Earth, but where was the proof?

Francis Crick, co-discoverer of the double helix, captured Woese's imagination early in his career. Crick wrote in 1958: "Biologists should realize that before long we shall have a subject which might be called 'protein taxonomy'—the study of amino acid sequences of proteins of an organism and the comparison of them between species. It can be argued that these sequences are the most delicate expression possible of the phenotype of an organism and that vast amounts of evolutionary information may be hidden within."

Woese was apparently the only scientist listening because he was the only one who bothered to try to discover those hidden evolutionary secrets. There was something especially seductive to Woese about

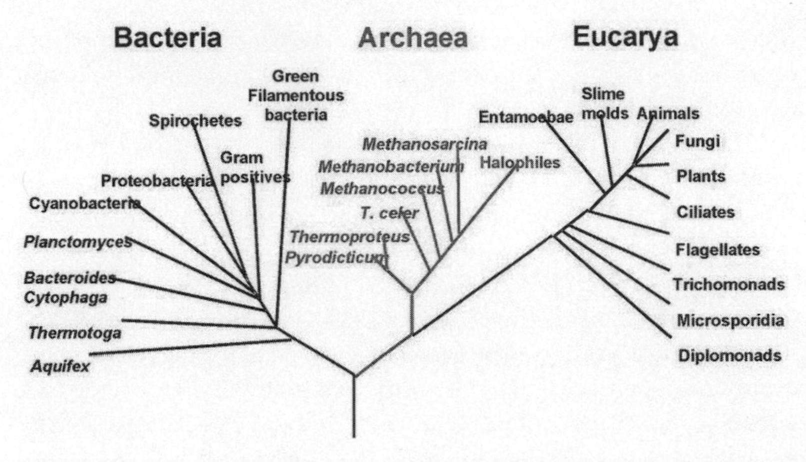

Carl Woese's tree of life. It depicts the archaea, bacteria, and eukaryotes. So far, all life on Earth belongs to one of these three main branches of the tree. *Courtesy of Carl Woese.*

fundamental principles, laws of nature. He confessed in an unusually personal interview published in *Current Biology* that he felt destined to become a scientist:

> I had to. No other way to cope with my world. Hard to explain though, because the child who made the decision was too young to verbalize it. There seemed to be two worlds, that of nature and that of people. The first was vast, wonderful, inscrutable, frightening, exciting, enticing, always moving, but nevertheless with an immutable consistency—it was a never-failing touchstone of truth. The world of people was the opposite: inconsistent, ever arbitrary, full of contradiction, anthropomorphizing, untrustworthy—almost devoid of truth. Growing up was a continual search for truth. In mathematics and science I finally found it; the "q.e.d." of geometry and Newton's Laws were like a warm shelter in a storm. Why I became a biologist is unclear, though. I had no scientific interest in plants and animals and took only one "bio" course, biochemistry in my senior year at college. However, a young instructor in physics named Bill Fairbank—who later went on to become a world class low temperature physicist—advised me not to go into physics, but into the exciting new field of biophysics, and to do so at Yale, whence he had just graduated. I followed that advice, and here I am.

In 1953, Woese received his Ph.D. from Yale in biophysics—the descriptive for what was to become molecular biology. It was the same year Watson and Crick announced their discovery of the double helix structure of DNA. It wasn't long before Woese's inquisitive mind began aching for a molecular taxonomy of bacteria and a more scientifically grounded explanation for the evolution of life.

"The genetic code of course was all the rage," Woese said during a relaxed interview from his summer home at Martha's Vineyard. After returning from Costa Rica, my own mind had been aching to learn more about Woese's research and his remarkable mind. I became aware of Woese's work at a 1996 press conference announcing the whole genome sequencing of the first archaea, *Methanococcus jannaschii.* The press conference was hosted by *Science* magazine and Craig Venter, who then headed The Institute for Genomic Research (TIGR). Venter's appearance at any press conference in the 1990s was guaranteed to draw throngs of reporters. Venter was regarded as a maverick who successfully bucked the establishment at the National Institutes of Health (NIH) in 1992 and set off on his own to sequence the whole genomes of microorganisms. Woese appeared at the press conference via satellite from a TV station in Urbana, Illinois. Woese sat silently for the most part with a somewhat irritated expression upon his face, justifiably so, as I came to understand.

The conversation in May 2003 in the surf with Eric Mathur and Steve Briggs inspired me to learn more about Woese and archaea. I called Jim Barlow, the science writer who had written the press releases on Woese's work for as long as I had been at my newspaper. Barlow sent me a package of Woese's scientific papers and press clippings collected on his work since the 1970s. Out of that big envelope spilled a treasure. I poured through the material at home alone at night. Midsummer, I arranged to speak with Woese. At first he seemed reticent. But when I begged to hear the full story of his discovery of the third domain of life—as it had become known by that time—he seemed pleased. A number of productive conversations followed.

Woese is a slightly built man now in his late 70s, still in possession of a razor-sharp mind, a dry wit, and a thick head of white hair. He is naturally shy, even a bit suspicious perhaps with people he doesn't know well. He answers personal questions reluctantly, but admits to enjoying jazz music and an occasional imported beer. He has a fondness for plaid shirts, khaki pants, and sneakers. An edge of resentment breaks through in his tone and words from time to time. He still gets angry when he looks back at the treatment he received for proposing a third domain of life. It's not the criticism that makes him angry. It was the outright dismissal of his work by people who never looked at his data. The meanness of the critics dampened the effectiveness of his laboratory and cost him precious time defending his work that could have been better spent on advancing the understanding of the origin of life. Had his work been recognized then for what it is now, his lab would have quickly become a Mecca for microbiologists and evolutionary biologists worldwide. He would have been a popular lecturer, begged to present his insights at top meetings around the world. It did not happen quite that way.

Woese is not an intimidating sort, except perhaps with his intellect. One gets the sense right away that small talk is a waste of his time and that the pain of the past is something he'd rather keep hidden behind his twinkling eyes. When the conversation turns back to science, Woese's mood considerably brightens. He returns to the scene of the early 1950s, explaining how the discovery of the structure of DNA triggered an explosion of thinking and research into deciphering the mysterious genetic code. With Watson and Crick's two historic publications in 1953 in the journal *Nature*, Woese became intrigued by what the code could reveal about biology's big questions and immediately began thinking about the origin of the genetic code.

After receiving his Ph.D., Woese focused his attention on RNA, which was DNA's harder-working and probably older sibling. But before continuing with Woese's story, let's have a closer look at what was revealed as Woese plunged headfirst into the RNA world.

Microbes deserve top billing in the story of DNA. In the early 20th century, debate raged over whether hereditary information was contained in proteins or DNA.[5] In 1928, Frederick Griffin determined that cellular material from a strain of infectious pneumococcus could be transferred into a harmless strain of bacteria making the bacteria pathogenic. In 1944, scientists Oswald Avery, Colin MacCleod, and Maclyn McCarty at Rockefeller University discovered that extracting DNA from a heat-killed strain of pneumococcus and mixing it in with a harmless strain was enough to make the harmless bacteria virulent. DNA appeared to be the key.

Building on the work, in 1950, Erwin Chargaff at the Columbia College of Physicians and Surgeons compared DNA in yeast and bacteria and quite remarkably found that the "relative amounts of the DNA bases—adenine (A), thymine (T), guanine (G), and cytosine (C)—did vary between species. But most importantly, he found that within a species the number of A's and T's and the number of C's and G's was always equal."[6]

Poor Chargaff did not realize that he had discovered the secret of the genetic code. According to Woese, Chargaff let the critical discovery of base pairing slip through his fingers only to be re-discovered by others or at least placed into a context that revealed its relevance years later. Woese said Chargaff deeply regretted not grasping the significance of the variation of sequence between species and the critical pairing of the DNA bases: A always pairing with T, and C always pairing with G. This knowledge is the foundation for understanding the genetic code of all life on Earth.

During this very early period, James Watson was working under Salvador Luria at the University of Indiana. Luria was fascinated with how viruses called "bacteriophages" were able to infect bacteria via a mixture of DNA and protein. The answer turned out to be RNA, which is involved in translating the genetic code of DNA and in serving as the codex for making amino acids, which become proteins. The story of how Watson and Crick arrived at the double helix is widely covered. But the point to remember is that Watson received a good grounding in RNA under Luria before moving on to Cambridge to begin tackling the structure of DNA with Francis Crick, who was a physicist by training. Scientists pursuing knowledge in the nascent field of DNA research possessed a rudimentary understanding that genes resided on chromosomes and that chromosomes were made of protein and DNA and perhaps some RNA. But the structure of DNA remained a great mystery. Linus Pauling proposed a triple helix structure, with which Watson and Crick also were experimenting. Numerous laboratories were closing in on the discovery of DNA's structure, each with different pieces of the puzzle in hand but no grasp of the complete picture. The great "aha!" moment was provided by an X-ray made by Rosalind Franklin, who worked in the laboratory with Watson and Crick. A bit of deviousness occurred in the manner by which Watson and Crick came upon Franklin's famous X-ray, called "photograph 51," which revealed the double helix structure of DNA. Watson and Crick had surreptitiously pillaged Franklin's files. On April 25, 1953, Watson and Crick published their paper in the scientific journal *Nature* describing the double helix structure of DNA.

The double helix refers to DNA being structured as two strands that coil like a spiral staircase, one going up and the other going down—anti-parallel. The steps of the staircase are made of the bases A, T, C, and G with AT always pairing with each other and CG likewise always pairing. These are called base pairs, and they are glued together by a hydrogen bond. DNA has the ability to make exact copies of itself whenever needed by the cell as it divides. When this occurs the hydrogen bond is cleaved and the two single strands of DNA separate. At this point, the cell forms bases of A, T, C, and G to pair back up with the bases on the two separated strands to form two new, identical, double-stranded DNA molecules.

But DNA can't do this by itself. It needs the normally single-stranded sibling, RNA, to do the translating of the genetic code to make sure the pairing takes place properly. RNA gets its name from the sugar group called ribose. RNA and DNA both have a sugar-phosphate backbone with nucleotide bases attached to it. Like DNA, RNA contains the bases adenine, cytosine, and guanine, but instead of thymine, RNA's fourth nucleotide is the base uracil (U), which is a close chemical cousin.

In general, proteins do the chemical work of the cell by acting as enzymes that catalyze a cell's chemical reactions. DNA is made up of long sequences of the four nucleotides A, T, C, or G. RNA is made up of sequences of the four nucleotides A, C, G, or U. Proteins are sequences of amino acids, with 20 amino acids to choose from. The sequence of nucleotides in the DNA and RNA represents the sequence of amino acids from which a protein is formed. The nucleotides are taken three at a time in a redundant code to represent one of the 20 amino acids. A specific protein, called an RNA polymerase (along with many auxiliary proteins), opens the DNA double helix and transcribes it to messenger RNA (mRNA). mRNA is read by a large ribo-protein molecule called a ribosome. Ribosomes connect amino acids together into a chain creating proteins, rather like a child snapping beads together for a pop-bead necklace. The sequence of the beads (amino acids) is specified by the triplet nucleotide sequences in the mRNA.

Why DNA *and* RNA? Double-stranded DNA is very stable. This is why we can extract it from insects trapped in 40-million-year-old amber and from 40,000-year-old Neanderthal bones. RNA, however, is unstable. DNA must be stable because it stores the cell's instructional information—life's instruction manual. RNA is needed only when cells must use the DNA code to make cellular components and specific proteins. When the instructions have been transcribed, meaning the cell's work order is completed, the mRNA for the specific protein decays and the ribosomes stop making that protein. In general, the word "gene" denotes the sequence of nucleotides that specifies the sequence of amino acids in a given protein. So, most of the time gene = protein.

In the last 20 years we've come to know that RNA can also have enzymatic activity critical to cell function. For example, the active site of a ribosome, the part that actually hooks the amino acids together, is made up of RNA. So, today, gene means protein (most of the time) or RNA sequence (not as common, but essential for cell function). So we can speak of the amoA gene, which codes for a protein that converts ammonia to nitrite in soil, or, we can refer to the 16s rRNA gene, which scientists rely upon to create microbial phylogenetic trees.

Unlike the typically double-stranded DNA molecule, RNA is usually a single-stranded molecule. RNA is often the genetic material used in viruses. RNA can move around in the cells of living organisms and thus serves as a sort of genetic messenger. In eukaryotic cells, such as those in our bodies, RNA relays the information stored in the DNA of the cell nucleus to other parts of the cell where it is essential for making proteins.

In 1954, Russian physicist George Gamow hypothesized that a three-letter code would be involved in generating each of the 20 amino acids, and organized an elite group of scientists, including Watson and Crick, into what was known as the RNA Tie Club. In the 1950s, RNA was hot, becoming hotter, and the members of the RNA Tie Club were the coolest of the cool. The club was to have 20 members, each representing the 20 known amino acids. Their objective was to crack the genetic code. Each member was to receive a necktie with the double helix structure and a lapel pin with an abbreviation of one of the 20 amino acids. When writing to each other, the club members sometimes used their amino acid "nicknames." Watson was PRO, for proline. Gamow was ALA, for alanine, and Crick was TYR for tyrosine. The lapel pins did not represent the RNA code, however. Nothing substantial was known about the actual three-letter "codon" of RNA and DNA until 1961. A codon is defined as the three bases of a DNA or RNA sequence, which specify a single amino acid.

"As originally conceived by Gamow the genetic code was the cryptographic aspect of a fundamental biology problem," Woese said. "The codons were seen as physically templating their corresponding amino acids. Thus to know the codon assignments was to understand the physical/chemical interactions upon which translation and its evolution were based."

Translation refers to the process in cells of reading the genetic code on DNA and making amino acids and proteins based on the code. Several teams of scientists in the United States and Europe attempted to work out the relationships among RNA, DNA, and protein, especially what were the genetic instructions and steps that led cells to manufacture proteins. James Watson, Francis Crick, Sydney Brenner, Har Gobind Khorana, and Dieter Söll were among the elite trying to solve this mystery. Joshua Lederberg, who was not a member of the Tie Club, also was interested in determining the code.

But it was Marshall Nirenberg, who had just completed a two-year postdoctoral fellowship at NIH in 1959, who was the first to describe a codon in 1961. Like others, Nirenberg suspected that RNA might be the messenger—the link between the genetic code of DNA and the manufacturing of proteins.

The codon and mRNA were both discovered in 1961. Nirenberg's education was in zoology and biochemistry with a focus on the biochemistry of tumor cells. He had no formal training in molecular genetics, which made the elite boys consider Nirenberg little more than a novice stepping onto other, more qualified, scientists' turf. According to the Nirenberg Papers at the National Library of Medicine, he had attended only evening courses on genetics, intended for intramural scientists at NIH.

"Many of Nirenberg's colleagues felt that it was naive for a biochemist untrained in molecular genetics to commence a brand-new area of research; at least one felt that Nirenberg was committing 'professional suicide.' In 1960, Nirenberg was joined by Heinrich J. Matthaei, a postdoctoral researcher from the University of Bonn in Germany then studying at Cornell University."

Nirenberg and Matthaei studied the DNA nucleotides—adenosine, cytosine, guanine, and thymine, and, of critical importance, they decided also to look at the role of uracil, which in RNA takes the place of thymine. Based on work by the Swiss geneticist Alfred Tissieres, Nirenberg and Matthaei created a synthetic RNA molecule and inserted it into *E. coli*, which was the laboratory workhorse for genetic researchers. As history records it, Nirenberg received most of the credit. But in fact, Matthaei did much of the work using an enzyme—the Ochoa enzyme—named for Severo Ochoa who received a Nobel Prize in 1959 for the enzymatic synthesis of ribonucleic acid.

According to the Nirenberg Papers, "On the RNA strand, synthetic RNA made of multiple batches of three units of uracil directed an amino acid chain composed entirely of phenylalanine. One three-unit batch of uracil could be read as UUU (poly-U), which was a three-letter shorthand method or 'code word' for identifying phenylalanine."[7] This reportedly was the messenger the scientists had been looking for. "The experiments showed that 'messenger RNA,' which transcribes genetic information from DNA, directs protein synthesis. That is, messenger RNA transmits the DNA messages that prescribe the assembly of amino acids into the complex proteins that drive living processes."

In August 1961, Nirenberg and Matthaei published their work in the *Proceedings of the National Academy of Sciences*, and then Nirenberg gave his first lecture on the U-3 discovery—as it came to be known—in Moscow to a small group of about 30 scientists at a session of the International Congress of Biochemistry. Francis Crick was one of the scientists attending the session. Crick immediately realized the significance of the work and arranged for Nirenberg to present the paper to the entire Congress of about a thousand people. Nirenberg became an instant scientific celebrity.

"By January 1962, interviews and photographs featuring Nirenberg and Matthaei appeared in scientific journals, newspapers, and weekly magazines around the world. UUU was described as the first word in the chemical dictionary of life, and the key to deciphering the entire genetic code," the Nirenberg Papers explain.

According to the Papers, "Nirenberg's work catapulted the scientist—whom the *Washington Post* described as 'painfully modest'—to international fame." James F. Hogg, Nirenberg's former advisor at the University of Michigan, joked in a letter: "In view of the very extensive recent publicity, we are considering putting a sign on our house, as follows 'Painted by Marshall Nirenberg' A.D. 1953. Would you please send a letter of authentication? We could then perhaps obtain a tax exemption as a historical site!"

At the end of 1961, Crick, Brenner, L. Barnett, and R. J. Watts-Tobin published a paper titled "General Nature of the Genetic Code for Proteins" in *Nature.* This group confirmed that the genetic code consists of the three-letter codons.

In his autobiography Brenner wrote, "In late 1962, Francis Crick and I began a long series of conversations about the next steps to be taken in our research. Both of us felt very strongly that most of the classical problems of molecular biology had been solved and that the future lay in tackling more complex biological problems. I remember that we decided against working on animal viruses, on the structure of ribosomes, on membranes, and other similar trivial problems in molecular biology. I had come to believe that most of molecular biology had become inevitable and that, as I put it in a draft paper, 'we must move on to other problems of biology which are new, mysterious and exciting.'"

Brenner is among the first to acknowledge the hubris of his declaration. It is safe to say that humility was a term rarely associated with scientists during the remarkable period of discovery of DNA and RNA in the 1950s and early 1960s.

Francis Crick also recalled the excitement and arrogance of the time in his 1988 book *What Mad Pursuit*:

> In his typical way, [Gamow] had founded that unusual organisation, the RNA Tie Club. There were to be only twenty members, one for each amino acid, and not only did each member receive a tie, made to Gamow's design by a haberdasher in Los Angeles, but also a tie pin with the short form of his own amino acid on it. I think I was Tyr but I'm not sure I ever got the tie pin. As it turned out the club served as a mechanism for circulating speculative manuscripts to the few people interested. . . . The main idea was that it was very difficult to consider how DNA or RNA, in any conceivable form, could provide a direct template for the side-chains of the twenty standard amino acids. What any structure was likely to have was a specific pattern of atomic groups that could form hydrogen bonds. I therefore proposed a theory in which there were twenty adaptors (one for each amino acid), together with twenty special enzymes. Each enzyme would join one particular amino acid to its own special adaptor. This combination would then diffuse to the RNA template. An adaptor molecule could fit in only those places on the nucleic acid template where it could form the necessary hydrogen bonds to hold it in place. Sitting there, it would have carried its amino acid to just the right place it was needed. . . . The paper was circulated to members of the RNA Tie Club but was never published in a proper journal. It is my most influential unpublished paper. Eventually I did publish a short remark briefly outlining the idea and tentatively suggesting that the adaptor might be a small piece of nucleic acid. . . . As every molecular biologist now knows, the job is done by a family of molecules now called transfer RNA.

Watson reflected on his first introduction to RNA in his book *Passion for DNA*: "RNA first came alive to me during the fall of 1947 at Indiana University when I took Salvador Luria's course on viruses. . . .

Apparently a given virus had either RNA or DNA in contrast to cells which contained both."

While Watson and Crick were becoming scientific legends in the 1950s, Salvador Luria's interest in RNA was building his own claim to fame for his research on phages, a type of virus that infects bacteria. One of Luria's research passions was to find a living system that was as simple as possible to help him discover fundamental life processes. Topping the list of things to do for Luria was learning the secret of self-replication—which is essential for the origin of life. RNA, which has the ability to self-replicate and is necessary for DNA to replicate as well, was the best enabler for Luria's goals.

Luria and colleagues Max Delbrück and Alfred D. Hershey shared the 1969 Nobel Prize in Medicine for their seminal work on the bacteriophage. According to the Nobel Prize Committee, "They studied what happened in single cells and analyzed their results with advanced statistical methods. They made a series of fundamental discoveries. . . . Bacteriophages have served and continue to serve as models for the more complicated and less approachable systems represented by animal and human cells. Delbrück, Hershey and Luria have set the solid foundations on which modern molecular biology rests. Without their contributions the explosive development of this field would have been hardly possible."

After Nirenberg, Crick, Brenner, Khorana, Söll, and the others published the seminal work on life being "a three-letter word" as Sagan put it, debate rose quickly about the actual significance of the RNA code. Gamow theorized that the codons that corresponded with specific amino acids were absolute and somehow meaningful. But Crick incorrectly believed the codons were arbitrary and whichever codon had become associated with an amino acid was just a historical accident.

Crick's hypothesis on the evolution of the genetic code held that if the RNA genetic code were to evolve again an entirely different set of codons would randomly find their assignments to their amino acids. Crick's assumption that codons were arbitrary became almost universally accepted at the time although it was later seen to be incorrect.

Nirenberg played a key role in overturning Crick's hyphothesis on how codons were assigned to amino acids. Nirenberg, Khorana at the University of Wisconsin, and Robert Holley at Cornell University, shared the Nobel Prize in Physiology or Medicine in 1968 for "their interpretation of the genetic code and its function in protein synthesis." Holley focused on alanine transfer RNA. He showed that tRNA is necessary for building proteins from amino acids. Söll, who is currently at Yale University, emphasizes that this early work on tRNA actually laid the foundation for the development of genetic sequencing.

Every living organism contains ribosomes, which are encoded by genes with similar sequences of the four letters of the RNA alphabet regardless of species. Nature is known to be very conservative when it comes to genes that carry out fundamental processes in the cell. If a gene is working fine for instructing a cell to manufacture a particular protein in a microbe, nature will use the same gene with only slight variations in the order of its code to perform the same task in other species of microbes as well as in humans. This phenomenon is called the "conservation of genes." Scientists will say a gene is "highly conserved across species," meaning the same gene does the same thing in different organisms. The suite of genes that instruct cells to form wings in a fruit fly, for example, is essentially the same as the suite of genes that instruct human embryos to form limbs. The genes of the ribosome are among the most highly conserved in the genetic code of all life forms because RNA is so fundamental to translating the genetic code of DNA into the basic components of cells and thus of life itself. Because of its critical importance to basic cellular function, RNA is uniquely resistant to random mutations. Its genetic code does not vary by much between different species.

And this is where Woese enters the RNA research story. To Woese, this suggested that the coding of ribosomes held grand promise for revealing fundamental evolutionary information about cells. Woese theorized that RNA in the ribosome (rRNA) would change more slowly, potentially allowing evolutionary paths to be traced over greater periods of time. This was an incredible insight, which would lead him to become the first scientist to discover a method of truly classifying microorganisms.

He published papers in 1962 and 1963 arguing that the RNA code contained a pattern that could not be arbitrary. While Brenner and Crick had assumed the big work was done, Woese intuitively felt they were missing the entire point. In 1965, Woese stressed again in a third seminal paper:

> While it is important to know what the genetic code codon assignments are, it is more important to know why they are, i.e., to know the mechanisms giving rise to the particular assignments observed. Only when

> the latter question is answered can we truly claim to begin to understand the genetic code. To date, however, most scientific attention has been turned to the former question. The reasons for this are quite clear. For one, the experimental systems for answering what the codon assignments are have been developed and are rapidly yielding a solution, while those for elucidating why such codon assignments exist are either not yet discovered or are very far from yielding the sought-for answer. For another reason, [Crick's hypothesis on the evolution of the code], one of the central dogmas of the coding field today, predicts the answer to the "why" question to be entirely uninteresting and trivial.[8]

Woese has always searched for the "why" in every endeavor. At the time of the interview from Martha's Vineyard, Woese was finishing a manuscript he proudly referred to several times during our talk. It was to be a "Personal Perspective" titled "The Archaeal Concept and the World It Lives In: A Retrospective."

Woese wrote: "From the start I had been skeptical of [Crick's theory]. The idea never really explained or predicted anything; we were just as ignorant with as without it. What [the idea] did do was spread a veneer of complacency over our ignorance—which of course silenced all further questioning" concerning transfer RNA, or tRNA. Woese saw something fundamentally important in tRNA that the elite RNA Tie Club completely missed. Watson appears to have missed the significance entirely too. When Watson first heard of the early work on transfer RNA he scoffed, Woese said.

"It did not feel right that the code was merely an historical accident. There was too much order in the set of codon assignments for that; order which had to be explained. The problem of the genetic code should never have been formulated in a vacuum in the first place. The code was manifest materially in the tRNAs, and to understand how the code evolved, one could not simply ignore the evolution of the tRNAs, or the rest of the translational apparatus for that matter. Right or wrong this is the argument that convinced me that the central problem in the evolution of the cell was the evolution of translation, and from then on this has been my major, driving concern in biology," Woese wrote with satisfaction.

However, it is one thing to be right, and another to be part of the in crowd, and Woese wasn't, even though he received his Ph.D. at Yale, worked at the Pasteur Institute, and landed a job at GE's main research laboratory in Schenectady, New York. After leaving GE in 1964, Woese set up his own laboratory at the University of Illinois and got down to the business of determining the significance of the RNA code himself.

One strike against him was that he was neither a microbiologist nor an evolutionary biologist. Taxonomy was not a playground for biophysicists. Plus, big-time microbiologists admitted in the early 1960s that organizing bacteria into a sensible phylogeny was impossible. Roger Stanier wrote an article in 1962 proclaiming the impossibility of classifying microbes. I should also note here that the well-regarded 1963 textbook by Stanier and other notable microbiologists makes NO mention of anything called a prokaryote when discussing the classification of life.

In 1963, Stanier and colleagues M. Doudoroff and E. A. Adelberg wrote in the second edition of *The Microbial World*: "the abiding intellectual scandal of bacteriology has been the absence of a clear concept of a bacterium."

Woese likes to quote that passage. He says after failing to create a taxonomy based on shape and other physical characteristics of bacteria, the big boys slipped in the prokaryote definition, which conveniently became a wastebasket in which microbiologists could dump everything that they couldn't understand.

Science writer Virginia Morell captured the resignation leading microbiologists felt about classifying bacteria. In her May 2, 1997, article on Woese for *Science*, "Microbiology's Scarred Revolutionary," Morell wrote that C. B. van Niel of Stanford University's Hopkins Marine Station declared in the late 1930s "the classification of bacteria was the key unresolved issue of microbiology." But after decades of fruitless labor trying to classify bacteria by their shape and metabolism, he and his former student Stanier, of the University of California at Berkeley, decided that bacteria simply could not be phylogenetically ordered. "The ultimate scientific goal of biological classification cannot be achieved in the case of bacteria," Stanier stated in *The Microbial World.*

Woese told Morell, "I hadn't been trained as a microbiologist, so I didn't have this bias." She wrote that his "physics background led him to believe that 'the world has deep and simple principles, and that if you look at it in the right way' you can find these."

Good thing Woese was an independent thinker! The tools for creating a molecular tree of life did not exist. Brenner and Crick declared they had solved the problems of the genetic code and determined that ribosomes were uninteresting. The leading scientists in bacteriology had stated a phylogenetic tree of microbes was impossible. No wonder Woese was the ONLY scientist who believed the ribosomes possessed evolutionary secrets and that a tree could be developed and must be developed for microorganisms. When Woese set up his lab at the University of Illinois, the time was ripe to pioneer a scientific revolution.

"The story begins back when I was at General Electric in the early sixties, from '60–'64. I was at the Knolls Research Lab in Schenectady, New York," Woese said with a manner that tells you to settle in. "I got there and of course didn't have any equipment."

That is an understatement. Woese was given Irving Langmuir's cloud-seeding lab. It was a bit outdated. Langmuir had been there from 1909 to 1950.

"But sequencing of proteins was just beginning to take off and you could see that there was evolutionary information in the proteins. Even Crick had thought so, but blinded perhaps by the light of all the halos surrounding the elite, he couldn't see where to look," Woese said.

"I had worked previously on ribosomes. There was by this time a front-page cover on the genetic code in a prominent journal from a theoretical point of view. By then Gamow had formed the Gamow Tie Club and they had 20 ties, and they were the elite of who studied the genetic code. This was a short-lived period. The theoretical work they were doing wasn't in touch with reality. I was caught up in the excitement and was working on the code from a Mickey Mouse theoretical point of view. But I began to realize that what we needed to be talking about was the evolution of translation, a whole set of problems above Darwin. Only when we moved into the molecular level could we get any handle on these kinds of questions. So I began looking at the intersection between proteins and amino acids. Then after 1964, I moved to Illinois where I have been ever since."

Once he was settled in at the University of Illinois at Urbana-Champaign, Woese, age 36, could not rest until he pursued the deeper evolutionary meaning in ribosomal RNA. He had selected this midwestern university because it gave him the freedom to follow an unproven path, his own path. For Woese, truth about life's origins and the origin of the genetic code would be revealed only through a molecular perspective of the microbial universe.

"To be serious about studying the genetic code we would have to have a universal framework. My number one job was to create the framework. People were putzing around with plants and animals which phylogenetically were just a couple of twigs on the tree of life. I had the molecular spirit in me and was set to determine the evolutionary tree, determining the evolutionary problems from a post-Darwinian level," Woese said at Martha's Vineyard.

While Woese was ramping up his laboratory, Frederick Sanger and his colleagues at Cambridge published a paper in 1965 on their new method of gene sequencing. It was called "RNA Oligonucleotide Cataloguing." It did not allow for complete RNA/nucleic acid sequencing but it gave Woese the tools he needed to begin determining a tree of life based on the RNA code.

This oligonucleotide was a short stretch of RNA, typically in the range of 20 letters or less, using the RNA alphabet of A, C, U, and G. Woese used enzymes to cut the segments into smaller pieces and then used Sanger's initial sequencing method to determine the code. The oligonucleotides appeared as small rows of black shadows on a large sheet of photographic film placed over a light box. The individual shadows represented letters of the alphabet and were of varying lengths according to whichever base they represented. The net result looks a bit like a blurry barcode.

"Fred Sanger is probably the greatest biologist of the 20th century because he pioneered the sequencing of proteins in the early 1950s when no one else saw the critical importance of doing so," Woese said. "And then he went right to work on the sequencing of nucleic acid and they all wrote him off, again not recognizing the importance of the endeavor until he knocked them over a quarter of a century later with his dideoxy sequencing method for DNA!"

The dideoxy method is what made sequencing the human genome possible. Self-centered by our nature, most human scientists in the 1960s focused their attention on human genes. The bacterial systems—the *E. coli*—scientists worked with were merely vehicles for getting at the expression and sequencing of human genes. Woese, however, went straight to sequencing RNA in microorganisms for the deepest possible evolutionary insights into all life. Nearly all of the wall space of his lab became covered with light boxes displaying films of RNA coding of bacteria. From the beginning, Woese had a hunch that ribosomal RNA held the secrets to the tree of life.

Morell wrote: "Physically, these RNA fragments appeared as fuzzy spots on film, and Woese stored thousands of such films in large canary-yellow Kodak boxes. It took him a full year to make and read his first catalogue. He was one of 'only two or three people in the world' to learn this backbreaking technique, he says, and labored alone. 'Carl was the only one who could read the films,'" recalls William Whitman, a former postdoctoral student of Ralph S. Wolfe, also at the University of Illinois.

Woese told Morell that the work was so tedious that it left him "just completely dumbed down."

Woese focused his attention on 16s rRNA genes, which he had known from previous work to be highly conserved across all bacterial species. Ribosomes are essential to life. The 16s rRNA gene was chosen for sequencing because every microorganism that contains a ribosome possesses a version of the gene. Using his new technique, Woese was able to determine about 25% of the sequence order of the 16s rRNA gene of thousands of individual bacteria representing hundreds of species and subspecies. He determined that the more similar the sequences are between microorganisms, the more closely they are related. Likewise, greater differences in the sequence order reveal more distant species relationships. This method allowed Woese to at last begin constructing an architectural schematic of the microbial universe. Branch by branch, his new tree began to grow. This molecular taxonomy existed nowhere else but in Woese's lab.

"It turned out that reworking Sanger's oligonucleotide cataloguing technology to fit our needs was the least of the problems we would face. At the time I was totally unaware of the hornet's nest we were stepping into with bacterial taxonomy—something that had a long and sorry history," Woese wrote in his Personal Perspective. "To continue the story: the oligonucleotide catalogue method was soon ready to go, and we began (at first very slowly) to turn out rRNA catalogues (sets of characteristic oligonucleotides) for a variety of microorganisms and a few eukaryotes. In the process, one quickly gained a feel for whether a particular catalogue represented a 'prokaryote' or a 'eukaryote' . . . each grouping had a distinctive 'oligonucleotide signature.'"

Up to this time, the microbiology community had not paid much attention to Woese's work. Those who were invited to look at the films plastered over his light boxes simply gaped in astonished ignorance.

"In 1976, I had been going after this for at least six years, ramping up and getting a little better with my technique. No one outside of us knew what was there," Woese says a bit sardonically. But Woese's lab was just down the hall from Ralph S. Wolfe's, and Wolfe was the person who introduced Woese to the methane-fuming microbes cultured from the sewage sludge. The two men spoke often about each other's work.

Wolfe supplied much moral support for Woese, who toiled alone. They were beginning to form their own society.

At the time, microbiologists knew that methane-producing bacteria lived in the intestines of cows and could be found in oxygen-deprived swamp sediments and sewage sludge. Robert Hungate was the sage of the day on these peculiar anaerobic organisms. Hungate is regarded by many as the first modern microbial ecologist.[9] Hungate was president of the American Society for Microbiology in the early 1970s. He had a wonderful academic pedigree. Hungate was a student in the first microbiology course taught in 1931 by C. B. van Niel at the Hopkins Marine Station of Stanford at Pacific Grove, California. Remember the marine biologist "Doc" in John Steinbeck's *Cannery Row*? The character was based on Ed Ricketts whose lab was also at the Hopkins Marine Station along with van Niel's on Cannery Row. While Steinbeck was inspired by Ricketts, Hungate was inspired by his young professor van Niel, who became the leading microbiologist of the first half of the 20th century. Hungate went on to pioneer the culturing of methane-producing microbes and developed the first laboratory techniques to grow them in anaerobic cultures. Hungate's source of methanogens was the rumen of cows. Hungate studied the rumen as an ecosystem, which revealed information about how the microbes made a living in the rumen. In 1966, Hungate published *The Rumen and Its Microbes.* This may not sound like a best-seller, but Wolfe became awesomely inspired by Hungate's book and research on the methane-producing cow belly bugs. Wolfe would end up at the University of Illinois down the hall from Woese, refining Hungate's methods of growing the microbes in oxygen-free jars with just the right nutrients.

As Wolfe saw Woese achieving greater proficiency with the molecular taxonomy of bacteria, Wolfe became curious about where the oddball methane-producing bacteria of his cow manure might fit on Woese's ever expanding family tree. And he began to wonder about Thomas Brock's microbes encountered in Brock's Yellowstone hot springs investigations. Brock determined that the microorganisms found in salt marshes called halophiles and the sulfur-loving, heat-thriving bacteria were most likely rare species of bacteria that had merely adapted to their more extreme environments. They were regarded as a scientific curiosity. Even so, they preferred environments that were primordial, more similar to those that might have been common on the young planet Earth. Methanogens had not been studied much at all, perhaps due to a natural aversion to collecting samples from sewer plants and cow piles. But they too might have a deeper ancestral lineage. Wolfe happened to be one of the few scientists to take a keen interest in the methanogens. What incredible serendipity!

"Methanogens were being studied in Ralph Wolfe's lab. He had devised a very neat method for growing them anaerobically," Woese continued from Martha's Vineyard. "He would take a serum bottle, inoculate the organisms in it, pressurize it with the right amount of CO_2 and hydrogen, and let it sit, the whole thing being otherwise sterile, of course. Then the methanogens would grow in the culture. The University of Illinois is a land grant institution so we had an agriculture division, and we could go collect our samples from the local sewage plant and barns. We did not know at that point that methanogens grew at high temperatures in volcanic vents. But later, Tom Langworthy, a student of Tom Brock, gave us cultures of *Thermoplasma* and *Sulfolobus.* We got halophiles from a collection at Woods Hole. They didn't look like anything we had seen either," Woese said, pausing to let this sink in.

Woese and Wolfe had no idea in the beginning what they were looking at. The ribosomal sequences were not at all similar to any of the hundreds of bacterial species that Woese had so meticulously cataloged over the years. Perhaps the anaerobic sewer bugs, the salty dogs, and the sulfur-loving hot-tubbers did indeed represent an ancient lineage of bacteria, ancestors that perhaps lived on Earth before the planet developed its present atmosphere. Then again, maybe Woese made a mistake and the variations of the sequence reads simply were wrong. Woese did what every good scientist must do. He repeated the tedious oligonucleotide cataloging of the strange microorganisms.

"The first thing you think is you made a mistake in the experiment, so you do it over and to your surprise it comes out the same. Then you take another methanogen and do it, and lo and behold it has unique features too," he says.

Woese was extremely careful with his techniques because he knew the scientific community would scrutinize his methods once he dared claim what he was now considering. For that matter, they might not even understand his methods. Woese realized he would need more evidence. He began reaching out to others who were working on the unusual microbes for their own reasons.

"It had been known that the extreme halophiles, the thermoplasmas, and *Sulfolobus* all had the same kind of very unusual lipids: the links to glycerol were ether, not ester; the lipid chains were branched (built of isoprenoid subunits), not straight; the chirality around the central carbon of the glycerol moiety was the opposite of what one would expect of a typical lipid," Woese said. "What was not known at the time was what sort of lipids the methanogens had. Ralph Wolfe prepared a goodly amount of a methanogen and the cell mass was shipped to Tom Langworthy, who had finished his work at Brock's lab and was at the University of South Dakota. Langworthy analyzed the methanogen's lipids, and voila. They possessed the same unusual variety seen in the other weird microbes."

The descriptions of the composition and structure of cell walls will sound technical to most of us, but these are the same characteristics scientists would examine if microbes were discovered in a sample of soil returned from a Mars lander, or found in the center of a basketball-sized ball of comet ice found melting in a crater anywhere on Earth. Some of the differences that Woese and his colleagues noted between the lipids of the halophiles, *Thermoplasma* and *Sulfolobus*, and those of bacteria were described initially by Brock.

The lipids of the halophiles were first described by Morris Kates in Canada. Brock had noticed that his thermophiles had the same unusual types of lipids, but he missed the significance of those differences. Brock had attributed the ether bonds and branched chains of the fatty acids, as well as the opposite chirality, to convergent evolution—which means that different species can arrive at the same adaptive solution to similar environments. Brock's conclusion fit with that of the dominant group-think on Darwinian evolution. The dominant leader of this group was Ernst Mayr, who at the time was at Harvard and later was at the American Museum of Natural History.

Woese thought Brock was mistaken and that Mayr was too preoccupied with the "twigs of plants and animals" to see the relevance of a brand new molecular phylogeny.

"At this point it became apparent that we were dealing with a major organismal grouping, a collection of highly disparate phenotypes all of which appeared to have the same pedigree. Still caution was called for. Before going off on cloud nine about finding a whole new kingdom of organisms, I had to make sure the organisms have more in common than just a characteristic ribosomal RNA," Woese said.

"There was a prejudice that there were two domains of life. That bias had taken hold of biology. I had been working on as many bacteria as I could find, hunting around and asking as many microbiologists as I could find for their cultures. Right away you could tell from the pattern of oligonucleotides in the methanogen rRNAs that something was strange and wasn't typical of any of the other bacteria that we had done, or typical of any of the eukaryotes we had done," Woese said. "It turns out we were not wrong. It was the preconceptions that were wrong. I was startled by the finding, but only for a very short time. I remember going around and saying to people, 'What would something be if it isn't a prokaryote or a eukaryote?' They couldn't fathom that something could be neither one. They thought it might have to be something from Mars." Woese chuckled for a moment, but the sarcasm returned. "We were questioning the catechisms."

The scientists Woese invited to his laboratory to view his decade of exhausting work were skeptical or simply could not understand. They believed whatever Woese was displaying had to be a form of bacteria. What else could it possibly be? Woese was convinced that the microbes were not bacteria, but he remained wary of publishing his findings. The small society that Woese and Wolfe had formed in their labs was about to gain a truly powerful ally. He would prove to be lifesaving when Woese finally got the nerve to shake the branches and catch the attention of the dominant group. Wolfe had been corresponding with a highly respected microbiologist and botanist, Otto Kandler at the University of Munich. Kandler was very interested in the methanogens and was familiar with the thermal hot spring bacteria that Brock had been writing about from his lab. Kandler's collaborators, Wolfram Zillig and Karl Stetter from Germany, were independently discovering that the RNA polymerases used by bacteria cells to transcribe DNA code into messenger RNA were different from those of methanogens, sulfur microbes, and halophiles. They had found the methanogens also displayed insensitivity to an antibiotic that was common among the other "bacterial freaks." The pieces were coming together, but into what, Woese wasn't sure. The microorganisms still did not appear to be bacteria or eukaryotes. The prokaryote–eukaryote system that everyone accepted suddenly stopped making sense. Kandler was attending a conference in the United States and made arrangements to visit Wolfe in Urbana.

"George Fox [Woese's postdoctoral student] and I were prepared for the usual frustrating struggle to get the visitor to realize that there just might be something under the sun other than *E. coli* and company in the bacterial world. You can imagine our surprise when shortly after beginning our sales pitch Kandler's eyes widened and he said, 'Of course!' or something like that. He had an open mind and a great feel for biology. That, it turned out, was exactly what the archaea needed to push the concept into the mainstream. Kandler returned to Germany full of enthusiasm about archaebacteria and proceeded to use his considerable prestige and power in the German microbiology establishment to help German microbiologists see the validity and great potential in the archaebacteria."

Woese was 49 when he published two papers in succession on the strange microbes in the *Proceedings of the National Academy of Sciences (PNAS)*. The first included Wolfe and the methanogen lab. But the second and more important paper was by Woese and George Fox, proposing a new classification of life and a name for it. The historic paper was published in the November 1977 *PNAS*.

Imagine waking up in the morning, opening up your copy of the *New York Times* and seeing a front-page headline that reads: "Scientists Discover a Form of Life That Predates Higher Organisms." It would catch my attention. That headline ran in the *New York Times* on page 1 on November 2, 1977, written by Richard D. Lyons. The story began: "URBANA, Ill. Nov. 2—Scientists studying the evolution of primitive organisms reported today the existence of a separate form of life that is hard to find in nature. They described it as a 'third kingdom' of living material composed of ancestral cells that abhor oxygen, digest carbon dioxide and produce methane."

In Chicago readers were greeted on November 3 with the front-page headline in the *Chicago Sun-Times* that blazed, "Scientists Study a Third Form of Life." The article began: "In research with far-reaching evolutionary implications, scientists Wednesday proposed that an uncommon oxygen-hating micro-organism may represent a 'new' form of life with a line of descent older than any other creatures on earth."

The newspaper headlines naturally generated excitement. Johnny Carson, a big fan of Carl Sagan and astronomy, mentioned the discovery of new life in his monologue. NASA issued a joint press release with the National Science Foundation (NSF) with a headline that read,

"SCIENTISTS IDENTIFY ORGANISM AS POSSIBLE OLDEST FORM OF LIFE ON EARTH." It was in all caps. NASA and the NSF had funded part of Woese's work. The NASA news release went on to say, "'The discovery creates new hope that science will ultimately find out a great deal about how life came about on this planet,' Dr. Woese said. 'Also researchers will be in a better position to understand and discover life forms that may have evolved elsewhere in the solar system or beyond,' he added." Elsewhere in the solar system or beyond! NASA put that out there in 1977.

Newsweek followed up November 14, 1977, with an article by Peter Gwynne, "The Oldest Life." The article began, "All the biology texts in the world may have to be rewritten. As they have it, the earth has two forms of life, the lower, bacteria, and the higher, plants and animals. . . . Methanogens, as the organisms are called, have been known to scientists for about twenty years. But their appearance and size made them look like nothing more than somewhat unusual bacteria. The most striking thing is that they survive only in warm environments entirely free of oxygen—on the sea floor, in sewage, in the stomachs of cattle, or in the hot springs of Yellowstone National Park. They thrive by converting carbon dioxide and hydrogen into methane, the principle component of natural gas."

The interesting thing about these headlines and news articles is they were not cases of sensational journalism. The articles were spot on. The biology texts should have been rewritten 30 years ago. Woese nailed it. Woese had been working for nearly a decade on his molecular system of organizing microbes into a family tree. He had checked, double-checked, triple-checked his work and gathered even more supporting evidence. The archaea discovery represented old-fashioned diligence and the highest level of scientific integrity.

In Germany, news of the research had already been spread by Kandler to Zillig and Stetter. They wasted no time in following up on Woese's work.

Woese said, "Unbeknownst to me, another German scientist, Wolfram Zillig at the Max Planck Institute, a person who had spent most of his career on RNA polymerases, had heard of the archaebacteria, and had set about to examine their RNA polymerases. To his great surprise and pleasure, Wolfram discovered archaebacterial polymerases definitely not to be of the prokaryotic type. His career suddenly exploded into a great new adventure, both in the lab and around the world, where he and his young colleague Karl Stetter set out to find more 'bugs' in places that could be quite dangerous, like hot springs in Iceland, where both of them almost came to a catastrophic end on a couple of occasions. Stetter later continued this new swashbuckling brand of field microbiology—taking trips into the jungles of Thailand, darkest Africa, the bottom of the ocean, and other inhospitable places, to find new and exciting thermophilic archaebacteria. Everyone could feel their excitement."

Zing! Swashbuckling field microbiology; inhospitable places? Stetter's name kept popping up in conversations about archaea. Wherever archaea were, so was Stetter, which was no coincidence. I learned in Costa Rica, just two months before my visit with Woese, that Stetter was a co-founder of Diversa back in 1994 and that his rare collection of more than 45 species of archaea (at the time) was one of the early platforms in the company. I underlined Stetter's name in my notes with three exclamation points after it and scribbled "MUST CALL ASAP."

Woese's mood suddenly turned dark. Yes, archaea stimulated a tremendous amount of new research among German microbiologists. In fact, Germany became the world leader in archaea research, with the largest number of researchers in the field, the development of the most sophisticated technologies for studying archaea, the most prominent scientific journals devoted to the topic, and the largest number of papers published on archaea. But in 1977, the dominant group of Americans was not so receptive. Salvador Luria, the Nobel Prize winner in Medicine and pioneer of RNA research, poisoned the party. More than that, he and his cronies overturned the tables, broke the furniture, and made sure everyone ran away.

Woese said Luria had not read the archaea paper when it was published by *PNAS* in 1977. But Luria saw the headlines and reacted by deriding any notion of the possibility for the existence of a third domain of life. The hostility, Woese said, was shocking. Others soon followed suit, crossing boundaries of common courtesy by making fun of Woese. He was called a crank and a crackpot, out of his league being neither a microbiologist nor an evolutionist. Woese was a polite and old-fashioned sort of fellow. He anticipated criticism, but in the form of scientific debate. He welcomed challenges to his theory and was prepared to defend archaea at scientific conferences and in correspondences with other scientists in peer-reviewed journals. It was extremely difficult for him to fathom what was about to happen. He was not invited to conferences to speak about his work or defend it. Postdoctoral students did not flock to Woese's lab as young scientists were doing in Germany with Kandler and Zillig.

Woese had this to say about the experience:

> For many of us there could no longer be any doubt that the archaebacteria were a grouping of organisms unto themselves, neither eubacterial nor eukaryotic. That, however, is not how the majority of biologists, especially microbiologists, saw it. I was quite taken aback by the negative response the initial announcement of the existence of this "third form of life" evoked, and the vehemence of that response! My colleague Ralph Wolfe was telephoned by Salvador Luria, who scolded him: "Ralph, you must dissociate yourself from this nonsense, or you're going to ruin your career!" Two of the three main weekly news magazines in the United States carried a story about the discovery of the archaebacteria. The third did not—because, it turns out, the science writer for that magazine had checked with a microbiologist, a confidant of his, who had advised him that a "third form of life" was absurd.

This is common practice for science writers. When a dramatic new finding is made, we call the leaders in the field for comments. In this case, having a press release in hand from NASA and the NSF would lend substantial credibility to the story. Likewise, publication in a peer-reviewed journal tells us that a paper has passed muster at least among a handful of experts selected by editors of the journal to review and critique the manuscript before publication. Science writers also develop "confidants" whom we trust to help us understand and triage new information. Woese recalls the third national publication was possibly *US News and World Report.* Whomever the writer's confidant was, he or she must have been a heavyweight to quash a report that had the backing of NASA and the NSF.

"There was a notable amount of behind-the-scenes grumbling by microbiologists as well. But, strangely, only one biologist had the courage to challenge the archaeal concept in print at that time," Woese said. "The bizarre thing about this episode in the history of the archaea is that the grouping achieved notoriety not because they represented a third type of living system per se . . . but because their presumed existence violated a central dogma, the eukaryote–prokaryote dichotomy. So rather than question the dogma, most biologists were content to condemn the finding. Their failure to question that dogma is one of the black marks in microbiology's history, for the prokaryote–eukaryote dichotomy is a conjecture, a theory—and scientists are supposed to test theories. The history behind why microbiology did not do this is an interesting and instructive one, one well worth our considering."

Evolutionary biologist Ernst Mayr joined the attacks in 1990 when Woese began naming archaea species as formal taxonomic units in the scientific literature.[10]

"Until then, he had been very supportive of the concept. Naming them changed it all!" Woese said.

Mayr to his death refused to accept that archaea represent a new domain of life. Today there is still a lively debate. Sandra Blakeslee of the *New York Times* was one of the few newspaper reporters outside of Urbana to profile Woese in the 1990s. For her article headlined "In Defense of Bacteria: Scientist Devoted to Study, Appreciation of Microbes" she sought comment from Mayr. He told Blakeslee that Woese had gone too far with his molecular taxonomy and emphasis on microbes. "Woese would put dinosaurs and birds together," Mayr said. "While birds derived from something like dinosaurs, they have become so different that they should be recognized a different taxa in their own right."

Mayr was a great evolutionary biologist, but he missed the point entirely. He ignored the fact that well more than half of Earth's biomass is made of microbes and they had about a 3-billion-year head start on charismatic flora and fauna, which appear in the fossil record only 500 million years ago. In her article, Blakeslee quoted Woese's polar perspective, "It's clear to me that if you wiped all multicellular life forms off the face of the Earth, microbial life might shift a tiny bit. If microbial life were to disappear, that would be it—instant death for the planet."

Lynn Margulis of the University of Massachusetts—an evolutionary biologist and maverick in her own right—would agree with Woese about microbes sustaining Earth. She was a collaborator on the popular Gaia hypothesis, except she wasn't so fond of the warm-fuzzy Mother Earth reputation that the public began associating with Gaia. Her view was that Earth is not a living organism into itself. Margulis said, "An organism doesn't eat its own waste." Rather, the planet is an enormous continuous ecosystem comprised of smaller component ecosystems: an engine with interdependent parts. Driving the engine are the world's microbes.

Instead of a granola crunching Mother Earth, Margulis wrote in

The Third Culture: Beyond the Scientific Revolution: "Gaia is a tough bitch—a system that has worked for over three billion years without people. This planet's surface and its atmosphere and environment will continue to evolve long after people and prejudice are gone."

Despite the initial criticism and the successful campaign to ignore Woese, he persisted with his research, continuing to build evidence for the third domain throughout the 1980s and 1990s. His society of allies also grew larger and stronger. Norman Pace became a disciple of Woese's teachings. After the development of the polymerase chain reaction (PCR) in the 1980s, Pace began using Woese's 16s rRNA sequencing method for microbial ecology. Armed with PCR and more rapid methods of sequencing, Pace learned how to determine the identities and number of different species in any given environmental sample by amplifying 16s rRNA sequences. Using Woese's methods and insights, Pace further developed microbial ecology, spawning a wider academic pursuit of taking a census of both bacteria and archaea. This generated greater interest and funding from the Department of Energy (DOE). Woese's world of archaeal research paved a road that ultimately led to the founding of Diversa and other biotechnology companies interested in archaea's extreme biochemical pathways.

By the early 1990s archaea became widely accepted by microbiologists as a third domain of life, and scientists arrived at a new threshold with automated gene sequencing. With private funding Venter established TIGR and began collaborating with the DOE to sequence whole genomes of microorganisms. TIGR had sequenced a bacterial genome and then tackled a methanogen in 1996. The genome of *Methanococcus jannaschii* was compared to the bacterial genome and to sequences known from eukaryotes.

The DOE gave Venter at TIGR and Woese a joint grant to sequence a high-temperature deep sea methanogen. The sequencing was done at TIGR and Woese assisted in the interpretation of the data. The joint project proved once and for all that Woese was correct from the beginning in his assertion about archaea. When the project was complete, Venter and the editors at *Science* organized a press conference on August 25, 1996, to announce the finding that archaea was in fact a new domain of life. This was a very important day. But Woese does not like to travel unless it's to Martha's Vineyard! He decided against flying and thus attending the event at the National Press Club in Washington, DC. This was perhaps a big mistake. Venter was making headlines almost monthly. And while he is a good guy and extremely loyal to his friends, Venter believed that because he did the sequencing, he deserved credit for establishing that archaea was a new domain. Woese appeared on a video monitor via satellite from a TV studio in Urbana. It was Woese's second chance for widespread public recognition, but as they say in the theater, Venter stole the show. Venter was "hot" and whole genome sequencing of microbes was totally cutting edge.

"This brings to closure the question of whether archaea are separate and distinct life forms," said Venter at the press conference. "In decoding the genetic structure of archaea, we were astounded to find that two-thirds of the genes do not look like anything we've ever seen in biology before."

That two-thirds of the methanogen's genes were unknown to science was shocking. The reporters were all abuzz. As the media turned the spotlight on Venter and shouted questions at him at the National Press Club in Washington, Woese sat curiously still and quiet back at the studio in Urbana. Woese appeared uncomfortable and stiff on the large screen behind Venter. Reporters asked Woese only a few general questions. We were not given much opportunity.

"I considered it a command performance," Woese said. "I was sitting in the studio in Urbana before the press conference began and the phone rang. I was told it was from Washington. It was Venter calling. He said, 'I wanted to make sure you don't talk too much.'"

Venter says Woese misunderstood him and that he made every effort to share credit for the sequencing of archaea. "I didn't tell Woese to be silent. I asked him to prepare something short and sweet and be prepared to answer some questions. I have tremendous respect for Woese. I am absolutely in awe of someone who can come up with really good ideas based on a handful of data," Venter says. "Look, I did everything possible to make sure Woese was in Washington, DC, for the press conference. I offered to pay his way first class to fly him there." Those who know him, know that Woese can be prickly, especially after the criticism he had been through in the early days of his career. Venter says he never took credit for Woese's discovery, but that he does deserve credit for proving Woese was right. Over a dinner with me he asserts, "Keep in mind that *Methanococcus* was the third genome in history to be fully sequenced. Until then, most scientists still did not believe that archaea was a third domain of life. It took the genome sequence for everybody else, including myself, to catch up with Woese and really believe it." In fairness, Woese was quoted in numerous stories as being grateful to Venter for the sequencing.

Such is science. Woese and Venter remain quite cordial. Woese moved on from his single-purpose study of archaea and construction of a microbial tree of life to embrace the broader topic of the evolution of cells. "That tree which everyone wants to reroot, uproot, and chop down was only the appetizer needed to get on to the first course," Woese said.

He believes now that it may be impossible to determine whether archaea is the oldest branch of life because of a phenomenon discovered in the late 1990s. As more organisms were sequenced and their genes compared, it appeared that very early in the story of life a significant amount of opportunistic gene swapping took place between archaea and bacteria and eukaryotes. To get at the root of the tree, if it has one, Woese says we must now look at the evolution of cells per se.[11] Woese now theorizes that life did not begin with one primordial cell—the classic trunk or root of the tree. Instead, he proposes there may have been at least three simple types of loosely constructed cellular structures that swam in a pool of genes, evolving in a communal way, swapping genes or stealing them from each other as needed. Eventually these organisms became the three domains. Woese believes the driving force in the evolution of early life was this gene-swapping orgy known scientifically as "horizontal gene transfer." Refusing to back down shyly, Woese's new theory challenges the central tenet of evolutionary biology called the "Doctrine of Common Descent." The traditional theory holds that all life on Earth descended from one original primordial form. Woese's idea of communal evolution of cells has been germinating in his brilliant brain for decades. It is a wonderful theory: primordial life built with such a loose metabolic chassis that it can seize whatever energy is available under a wide variety of temperatures and chemical conditions. And he first published on this revolutionary idea as early as 1982!

"We cannot expect to explain cellular evolution if we stay locked in the classical Darwinian mode of thinking," Woese said recently. "The time has come for biology to go beyond the Doctrine of Common Descent."

Woese says that horizontal gene transfer has the power to transform entire genomes. With simple primitive organisms this process can "completely erase an organismal genealogical trace." In simpler terms, that means all of the early gene swapping completely tangled the roots of the tree of life. Whatever preceded archaea and bacteria may not have had a cellular form at all. But some form of metabolism would be necessary for life to arise. Woese imagines first life as loosely constructed, amorphous components swimming in a pool of genes. From that early, metabolizing ooze, cellular organisms might have evolved quite nicely.

Despite the damage done publicly to Woese and his lab by his critics in the 1970s and 1980s, he has collected some of science's highest awards. He received a "genius" research award in 1984 from the John D. and Catherine T. MacArthur Foundation. He was elected into the National Academy of Sciences in 1988. In 1992, Woese became the 12th recipient of microbiology's highest honor, the Leeuwenhoek Medal, given each decade by the Dutch Royal Academy of Science in the name of Antonie van Leeuwenhoek, the inventor of the microscope and discoverer of the microbial world. In 2003, he was awarded the $500,000 Crafoord Prize in Biosciences, which is given for accomplishments in scientific fields not covered by the Nobel Prizes in Science. This honor is as high as it gets, but Woese says he is most proud to see his tree of life used in nearly all biology textbooks. Now that's making history!

Blakeslee in her 1996 *New York Times* article captured a prescient quote from Mitchell Sogin, an evolutionary biologist at the Woods Hole Oceanographic Institution in Massachusetts: "If you pick up any biology textbook and look at the people who have been remembered you will see names like Darwin and Mendel. Tomorrow's biology textbooks will have names like Watson and Crick and Carl Woese. He works on the big picture instead of the little details."

Woese spent most of his career quietly taking the criticism as it came and publishing his rebuttals, usually with some subtlety, only in his scientific papers. So he surprised more than a few distinguished scientists on the evening of September 24, 2003, when he received the prestigious Crafoord Prize. After Woese walked slowly to the podium and began to address the King and Queen of Sweden and members of the Royal Swedish Academy of Sciences, barely a minute into his address to the audience, he let it fly. The following is most of the speech Woese gave to the Swedish Academy. Next to his marriage proposal, they are perhaps the most carefully chosen words of his lifetime:

"The Archaea were unexpected to begin with, and having arrived, they were unwelcome. This particular point needs understanding in that the situation is not unique to the Archaea."

Woese chose only one slide for his presentation. He displayed a diagram of the traditional two-domain tree of life on the upper half of the screen and his now established three-domain tree of life on the bottom half. Pointing to the slide he said:

> It depicts biology's knowledge of evolutionary relationships among organisms before our research program began and, then, a decade or so on into it. It is the background against which I now speak. The Archaea are biologically significant on several levels. Let me begin with the most superficial, the significance of the Archaea as an organismal group. From time to time amazing and important organisms are discovered. The tube worms that inhabit the environs of deep sea hydrothermal vents are one such find; the coelacanths, which resemble our Piscean ancestors, are another, as are bacteria that grow above the normal boiling point of water and fish that live below its freezing point. However, all such finds lie within the framework of established major taxonomic groups. They are like discovering major geological features on some continent.
>
> The discovery of the Archaea was different. It was more like discovering a whole new continent: the Archaea do not exist within any recognized higher biological taxon; they constitute a new highest level taxonomic unit in their own right. While they are microscopic, they have no specific relationship to the true bacteria, nor are they specifically related to eukaryotic cells. Thus, from a classical biological perspective, the Archaea are a truly major organismal find—the most significant one in modern scientific times. They have added a new trunk to the Tree of Life.
>
> I have already said that the Archaea did not fit into the 20th century biology paradigm. Let me expand on that, for it introduces the second level of their discovery's significance, the opening of the Pandora's Box of Microbiology. When first announced to the world the Archaea had been met with great public acclaim—for example front page coverage in the New York Times. That acclaim was matched only by the disdain accorded them by biologists in general, microbiologists in particular. The public had perceived their discovery as speaking to one of Mankind's eternal concerns—where we came from. Biologists received the discovery of the Archaea as anathema, which ran contrary to their conventional wisdom.
>
> You need to feel the intensity of biology's reaction to appreciate it: Our chief collaborator in the characterization of the methanogens, which were the first of the archaeal groups to be recognized, was Ralph Wolfe, my colleague at Illinois. When the initial press coverage of our discovery appeared, Ralph received a telephone call from a prominent Nobel laureate whom he knew well and greatly respected. His friend advised him in no uncertain terms to dissociate himself publicly from this scientific fraud or

his reputation would be ruined. Ralph, true scientist that he is, did nothing of the sort.

Unfortunately, their belief rested completely upon authoritative assertion, not scientific data, and, worse, an assertion whose truth had been proclaimed self-evident and undeniable. Schrödinger had said that the power of guesswork solutions is such that, in his words, "the answer is missed even when, by luck, it comes close at hand." The Archaea provided a perfect example. We were not specifically looking for them, merely trying to find out how various organisms fitted into some large-scale phylogenetic picture—as I said before. True to Schrödinger's characterization, microbiologists had indeed missed the answer, not this once, it turns out, but a number of times, for their guesswork solution had removed the question. A question answered is a question dismissed. Thus, the truth was not only overlooked, but when confronted with it, biologists flat out denied it. Like all such fabrications, guesswork solutions are vulnerable and require dedicated defense if they are to persist.

Thus, on what I have chosen to call its second level of significance, the discovery of the Archaea brought to light a flaw in the conceptual structure of Microbiology. Bacteriology, like zoology and botany, is fundamentally an organismal discipline; its focus is a group of organisms, which needs to be studied in four principal ways: First, in structure and function in order to learn how the organisms are built and work; second, in the group's diversity, to bring to bear the power of comparative analysis; third, in their ecology, to understand the role the organisms play in the tangled, interconnected web of life and, finally, in terms of their evolutionary relationships, to make biological sense of it all.

Meaningful development of the second and third of these turns upon the development of the fourth—evolutionary relationships. Otherwise the study of bacterial diversity amounts to no more than a catalog of disconnected vignettes about various microbiologists' favorite bacteria. And bacterial ecology exists in name only: how can you study ecology when you can't distinguish your "plants," as it were, from your "animals"? This is indeed the condition in which bacteriology found itself for the greater part of the 20th century: it possessed no phylogenetic framework, and as a result had no meaningful bacterial ecology or study of diversity. At base, Bacteriology was not a true organismal discipline, because bacteriologists had no real concept of the organisms they were studying. And their guesswork invocation of the 'prokaryote' had served only to obscure this all-important hiatus in their understanding.

Herein, then, lies the real temporal worth of our overall research program: It finally provided bacteriology with a phylogenetic framework within which to structure itself as a full-fledged organismal discipline. One can begin to see the salutary effects of this framework. Today talk of evolutionary relationships is no longer shunned, but is commonplace among microbiologists. Bacterial ecology has now come into its own and is fast becoming the dominant area of study in microbiology—just in time, I might add, because the problem of the global ecology, in which bacteria are absolutely central, is a pressing one.

In addressing bacterial ecology I would be remiss not to say that while a phylogenetic framework was necessary to the emergence of bacterial ecology; it alone wasn't sufficient. The vast majority of bacteria—and I mean well over 95%—cannot, for whatever reason, be readily isolated in laboratory culture. Norman Pace saw the way around this road block: given a molecular phylogenetic framework, he argued, meaningful identification of an organism requires only the characterization of some key gene, and that gene can be isolated from the organism regardless of whether it is in laboratory culture or cloaked in its natural setting. Pace's methodology for accomplishing this is what has made bacterial ecology feasible.

[Allow me to interrupt here to stress that Darwin did not even understand this point about a single gene tracing the whole genealogy. No classical biologists could have done so. It just slipped in with Woese's work.]

Finally, I come to the third, the deepest and most lasting, impact that our research program in general and the discovery of the Archaea in particular have had, namely helping to bring about the resurgence of interest in evolution. Throughout most of the past century evolution languished. By its nature the discipline of molecular biology had no interest in it. Evolution was simply a collection of idiosyncratic historical accidents, inexplicable and irrelevant to the understanding of biology. As we have seen, evolutionary considerations were absent from microbiology as well. And 20th-century classical evolutionists failed to press the frontier of their discipline into the molecular area. The net result was stasis. Once the universal phylogenetic tree appeared, however, it became difficult to ignore evolutionary relationships and evolution in general. What also became hard to ignore was that the molecular dissection of the cell was inadvertently providing the data needed to study how cellular organization evolved.

I had been wanting to attack the evolution of translation since the early days of the genetic code. But it was obvious that universal evolutionary problems could be approached only in the framework of a universal phylogenetic tree. And nothing even remotely approaching that tree existed in the 1960s. The microbial world was phylogenetic terra incognita. Consequently, I spent over two decades of my career determining that tree. The reward for me has not been primarily in seeing the tree per se emerge, for the tree is only the framework. The reward has been in seeing unfold the new and fascinating biological realm that lies buried within its roots.

It is in probing the evolutionary depth of the cell where the Archaea are proving their full worth. Because the Archaea provide a third example of basic cellular organization, biologists thereby have the equivalent of binocular vision: they can now "triangulate" on problems of cellular evolu-

> tion. And we all know how indispensable binocular discrimination is when the landscape is alien and complex.
>
> If one steps back from the day-to-day bustle of biology—from the frenzy of genomics and the race to diagnose, cure, and bioengineer our way to a utopia—then one can glimpse the new vision of Biology that is emerging. It is a deeper and more integrated vision of Biology than its molecular predecessor. I would like to feel that what we have presented to Biology and to the World—a universal phylogenetic framework and a third primary type of cellular organization—is helping to bring this new and exciting Biology into being.

Woese waited a long time to say all of that and he got considerable satisfaction in doing it. He was 75 years old when he addressed the Royal Swedish Academy. The University of Illinois has established a powerful new genomics institute at Urbana where Woese continues working on the evolution of cells and the origin of the genetic code. No living scientist has perhaps spent more time scientifically investigating the origin of life than Carl Woese.

When Woese was preparing his speech and steeling himself for the long flight to Sweden, I was at *USA Today* back on the science beat after returning from the Costa Rica trip with the Diversa guys. The afternoon standing in the surf with Mathur and Briggs kept drifting into my mind. I was aching to learn more about archaea. Speaking with Woese made me increasingly restless and curious. I called Mathur, relaying the great conversations with Woese over the summer. Mathur said he would be traveling to Yellowstone the first weekend in October to talk at a meeting of the Thermal Biology Institute—the American disciples who followed Woese's path and whose mantra was "16s rRNA archaea." Stetter, the German superstar, would be flying over from Germany as the keynote lecturer.

"You should come, man. It'll be great fun. Practically everybody who's working on archaea will be there," Mathur said. "After Yellowstone we're going sampling at a rainforest in Northwestern Kauai. Then Puerto Rico in February. We're diving for sponges there and we've been invited to Arecibo to meet with Jill Tarter, you know, the woman who runs SETI and was the person Jodie Foster played in Contact. How cool is that? And for sure we'll be going to Kamchatka next fall. You could see the whole archaeal world."

My heart was racing. Have you ever wondered what it would feel like to jump off a cliff without a parachute? After a month of sleepless nights, I told my editors at *USA Today* that I wanted to write a book about archaea. This idea was not met with enthusiasm. They had already given me time off to a write a previous book. The only options they offered were for me to do it in my spare time or to quit my job of 16 years. It felt a little sad, and a lot scary. I thought, "What the hell." Then I jumped.

4

THE WITCH'S GARDEN

Karl Stetter cultivates the world's largest collection of strange life.

Double, double toil and trouble;
Fire burn, and caldron bubble.
Macbeth, William Shakespeare

There he was—Karl Stetter, the "swashbuckling" daring do German microbe hunter, standing at the end of a corridor at the small airport in Bozeman, Montana. For more than 25 years he had been traipsing around the world, peering into live volcanoes, carefully choosing his path around hot springs and geysers, scuba diving on shallow volcanic vents, and riding submersibles to hydrothermal systems miles down to the bottom of the ocean.

He is tall and strongly built, wearing his substantial wealth, professional success, and intellect so comfortably it is not in the least glaring. A stranger would only sense something special about this guy, observe that he has good taste in casual clothes and luggage, and probably wonder why he carries two battered-looking shafts of bamboo pole duct-taped together. The poles obviously are not sections of a fly-fishing rod because at the tip of one a stainless steel cup has been firmly attached.

Eric Mathur of Diversa greeted me two hours earlier, crackling with enthusiasm about the beginning of our new adventure. He went to fetch the rental car, a new Ford Explorer, as I searched for Stetter. It was early October 2003, only one week since I left the newspaper. Fair to say I was in a bit of a daze; early freefall. The timing for this trip to the first Thermal Biology Institute conference at Yellowstone National Park couldn't have been better.

Karl Stetter's undying passion is to find the hottest of the hot and the strangest of the strange archaea. About 90 species of archaea have been identified since Woese's historic revelation in 1977. Stetter located and learned to culture more than 60 of them. In 2002, he discovered a bizarre form of archaea that holds the record as the smallest non-viral microorganism on Earth. Absent its own metabolism, it piggybacks on the outer membrane of its host and borrows genes for making energy—the ultimate freeloader. In 1980, Stetter discovered the then hottest organism alive, *Methanothermus fervidus,* living at 207 degrees F in boiling springs in the mountains of Iceland. In 1982, he topped that with *Pyrodictium occultum* (hidden fire network), which grows at temperatures up to 230 degrees under elevated pressure at a hydrothermal hot vent on the floor of the Mediterranean Sea at Vulcano Island. In 1986, he found *Pyrococcus furiosus*—the one Short and Mathur seized upon for a polymerase chain reaction (PCR) enzyme—growing at 220 degrees F.

In 1996, Stetter discovered the hottest of them all, *Pyrolobus fumarii* (firelobe of the chimney), which grows best at 235 degrees F. Stetter's discovery of so many organisms living at the upper limits of the outer limits led him to coin the term "hyperthermophiles" to describe them as a group.[1]

Thomas Brock paved the way for Stetter in the 1960s and 1970s. Brock, who focused on microbes at Yellowstone National Park, found the first thermophiles living at 150–170 degrees F, which nobody at the time had believed was possible.[2]

The hyperthermophiles are Stetter's microbial treasures. He found them in all the wild places of Earth, carried them in his aluminum briefcase back home to Germany, figured out whatever peculiar nutrients they needed, and then grew them up in ceramic and titanium cauldrons at his lab, which his German colleagues call the "witch's garden." He had to invent almost everything in the laboratory from scratch: the methods and the machines to grow organisms at pressures that will crush a car, temperatures that can boil humans, and acid levels that burn holes in stainless steel. Stetter's cultures are studied today by academic scientists all over the world. He is as proud of them as any doting parent and carries pictures of his archaea on his laptop.

Stetter looks much younger than a man in his early to mid-60s. I recognized him immediately from pictures Mathur e-mailed a few days before. I waved to him shyly. He had no idea who I was as I slung my backpack onto my shoulders and walked down the corridor to meet him. I explained that Mathur had invited me along and I was thinking of doing a book on archaea. Stetter smiled warmly and shook my hand with vigor. "Ah ha! That's good!"

We walked outside into a golden October sun. Mathur, waving eagerly at Stetter with one hand, zoomed up with the SUV and popped the hatch automatically before nearly screeching to a stop. Late afternoon was upon us, and soon the sun would dip behind the crest of the mountains. The plan was to spend the night at a small resort about a third of the way to Yellowstone. The resort had its own thermal spring and a fine restaurant. This was Mathur's way of taking care of his good friend who had flown from Munich to New York to Salt Lake and finally connected with the puddle jumper to Bozeman. By Stetter's watch it was already past midnight.

I sat in the backseat as Mathur and Stetter asked about each other's families and went over plans for the conference. Both were giving talks, but Stetter was the celebrity. He would present new data on an archaea species discovered at a volcanic vent at a depth of 350 feet on the mid-Atlantic ridge north of Iceland. Stetter named it *Nanoarchaeum equitans*, which means riding the fire sphere. (See Plate 3.) The initial discovery had been reported in 2002, and it had made big news. But its genome had just been sequenced in collaboration with Diversa, revealing much greater detail about the organism. Stetter was preparing to announce that his "nano" had the most compact genome of any known organism. There had been some dispute among scientists that the *Nanoarchaeum* was nothing new, only an unusual virus. But it was not a virus, Stetter assured Mathur. It was a parasite or a symbiont, neither of which had been seen in the archaeal domain before.

"Karl, that's really good news," Mathur replied. Diversa played a key role in the sequencing of the organism. The two men engaged in scientific small talk about the big discovery of the tiniest microbe on Earth. I listened quietly, gazing out at the landscape.

"So, Tim, you are from Washington?" Stetter asked shifting sideways in the front seat. "And this bad guy Eric has convinced you to write about the stink bugs and the other crazy archaea?"

Stetter refers to his sulfur-loving archaea as stink bugs, which he punctuates with a devilish grin. The man has spent nearly a third of his life with the scent of hydrogen sulfide in his nostrils. He associates the odor of rotten eggs with excitement and wonder. I related to Stetter the trip to Costa Rica, the conversations so far with Woese, the coverage of *Methanococcus* back in 1996, and talked a little of my general fascination with weird things. Stetter nodded his approval. "This is a very good thing then. You will come with us to take some samples, yeah? We will put you to work!" His friendliness and warm self-confidence evaporated the anxiety crawling up my back and burrowing into my brainstem over the past week.

The drive to Chico Hot Springs Resort took only an hour. Mathur drives at about the same speed he talks. We checked in and were shown to individual cabins, surprisingly luxurious. I kept forgetting these scientists are also successful entrepreneurs. Stetter excused himself to nap before dinner. I gladly used the time to scribble notes, and then met Stetter at his cabin at 8 p.m. Mathur arrived seconds later all smiles and enthusiasm. I've rarely met a more consistently upbeat person. Mathur and Stetter in the same room create a powerful synergy, and sense of mischief. We walked to the dining room, decorated with rustic paneling and stucco walls. The lighting was low; the tables set with warm candle-lit glass lanterns. Mathur selected a bottle of red wine, and the three of us ordered the same giant steaks. It was time to start work, so I asked Stetter how he came to be a microbiologist.

"As a little child I was very interested in microbes and so my parents gave me a little microscope, which I used to look at everything that was around. There were organisms in the rain water, which were of great interest of course to Leeuwenhoek when he was making his microscopes. I found these little bugs very interesting and would look at water from the ponds. But, also I was already very interested in cultivating plants," Stetter said as the wine was being poured.

Mathur stated that Stetter is an expert grower of rare orchids and, as it turns out, a very special grape. Mathur coaxed the secret from Stetter. Once at a fancy dinner party in an unnamed country with a vineyard dating back 2,000 years, Stetter was offered one of these rare grapes to taste. He kept the seeds of the grape in his mouth and when no one was looking placed them in a napkin and hid them in his pocket. Stetter has a greenhouse for the orchids at his home, then in Regensburg, Germany. The greenhouse now includes a grapevine with a remarkable, ancient provenance. Stetter lived in Regensburg and raised a family there since 1980 when he became a professor and chairman of microbiology at the University of Regensburg. He has three children. At his remodeled family home in Munich, he designed a much larger greenhouse that is a technical wonder. Stetter was raised in the same house.

"Then of course growing up I had an interest in airplanes and I thought maybe airplane engineering was the right thing. I had a friend who was in engineering at the technical university in Munich. I was also attending there. But, he was so good at his work this made me realize that the best I could become at engineering was mediocre, so after a few courses I had to think about this situation and do the right thing, to make the right decision about my future," Stetter said. "So I changed my subject to microbiology, and there was Otto Kandler."

Stetter's admiration of Kandler shines through his eyes. Kandler was the powerhouse of microbiology in Germany. Kandler had a burning desire like Woese to determine the taxonomy of bacteria. Kandler at the time was studying the chemical composition of cell walls and comparing the cell walls among different species of bacteria hoping to work out a phylogeny.

"It was true what everyone was saying that no one would ever find a phylogeny of microbes, because they are not like plants which have natural features. To find the relationships with microbes it was thought this is just impossible, but not for Otto. Therefore there were years of work he had been doing already on trying to make a chemotaxonomy. Then Otto realized the cell wall was a tool that could be used to trace the phylogeny of microbes," Stetter said.

Kandler was a gifted teacher, possessed with a mind always open to fresh and unconventional thinking. He saw in Stetter a kindred spirit and took him under his wing.

"Kandler knew that I was very interested in microbiology and he spent time with me to show me many bacteria. He had a lot of interest in cultivating bacteria and taxonomy, and so this was my beginning. He taught me many tricks about how to cultivate organisms, and I was able to isolate in my laboratory courses some new strains of bacteria. With this I became very excited," Stetter said.

Mathur smiled and nodded as Stetter talked. He was dying to jump in again. Stetter paused cautiously, allowing Mathur to interject that Stetter drives a Ferrari and loves to push it to 170 miles per hour on the autobahn. Cultivating organisms helped him to buy it.

"Karl is the consummate gardener of microbes," Mathur stated emphatically. Stetter waved his hand, embarrassed. He took a sip of wine, hiding his face behind his glass.

"Seriously, man. Karl invented a new way to make organic sauerkraut. It's sold in every health food store practically in Europe," Mathur said, pouring another glass of wine for us. The sauerkraut is marketed in Europe by a famous health food company. He explained:

> In my master's work I studied the formation of lactic acid, which is what lactic acid bacteria do. Humans are using lactic acids, lactobacilli, for thousands of years. They are used in making sauerkraut and yogurt, cheese, and in curing meat. They are in sourdough and kimchee—all fermented foods. Also they are naturally living in the small intestine of human beings to prevent pathogenic microbes from taking us over. Everywhere there were lactobacilli that could be cultured. You can find them also naturally in beet roots. Kandler was an expert on lactic acid bacteria, and we tested lactic acid production during the growth phase of the culture. In some strains the isomer changes during the growth phase. I learned to cultivate the lactobacilli and to purify a new enzyme. Then I had a new idea about how to make sauerkraut, and so we developed sauerkraut with only the L-isomer. This is because we found that the L-isomer is also made naturally in humans. In addition, we found the L-isomer tastes better.

"Making sauerkraut with a specifically defined microbe had never been done before. From this I developed a few new things that I was able to patent," Stetter explained with some lingering embarrassment. He took another sip of wine and paused, thinking about his next words.

"So, this was my first real money. You really can make money with microbes. And I believe this was one of the great moments of my career. I saw the first production of the sauerkraut and the labels coming off the production line, and I was touched. This was a very powerful moment and it led to the vision about Diversa. And this was a great experience for my lectures later on, because I could explain how to turn knowledge into a practical application," Stetter says. He set his glass down, leaned forward and with an earnest and serious tone he said firmly, "I am not ashamed to make money."

He and Mathur then burst into laughter and clinked their glasses.

In the beginning, Stetter was interested in studying the metabolism of bacteria. Working under Kandler at the University of Munich, Stetter studied the fermenting type of metabolism of the lactobacilli and the metabolisms, including reduction and oxidation reactions, of numerous other microorganisms. As he learned to culture microbes under Kandler's tutelage, Stetter also became familiar with the composition of bacterial cell walls. All of this provided Stetter with a unique combination of skills that would serve him well later when he began to study archaea.

Stetter's entrepreneurial spirit developed rapidly during his research for his master's degree and Ph.D. The lactobacilli served as the natural platform for Stetter's early research.

Stetter's second mentor, Wolfram Zillig, was an expert molecular biologist who studied transcription of the genetic code in *E. coli.* In particular, Zillig was a specialist in RNA polymerases. After moving on to Zillig's lab at the Max Planck Institute, Stetter began research on transcription in microorganisms, initially with lactobacilli. Transcription was of interest to Stetter because it is fundamental to all cells, whether human or microbial. During transcription, DNA sequences are copied via an RNA polymerase that transfers genetic instructions from genes on the DNA strands to the RNA strand. This is the first step that leads to translation of the RNA code into proteins. Carl Woese's interest in studying translation in many species of bacteria is what led him to select the 16s rRNA gene to develop his molecular taxonomy.

Stetter determined that transcription in lactobacilli was different from transcription in the *E. coli* in which Zillig specialized. *E. coli* was the platform for genetic research used by 99% of microbiologists until the late 1980s. It was essentially the only organism in which a genetic system was available for studying in the laboratory. *E. coli* were easy to grow and thus commonly used as the system for manipulating genes.

"Not many people knew about the transcription of lactic acid," Stetter continued. "After receiving my Ph.D. with Kandler I went to study under Wolfram Zillig because he was one of the world experts in transcription in *E. coli.* During this time I was working on my habilitation to become a professor and I finished in 1977," Stetter said.

Habilitation is an academic process in Germany and other European countries that involves three to five more years of independent research and teaching to qualify as a full professor. Under Zillig, Stetter had the opportunity to study transcription and RNA polymerases in lactobacilli.

As Stetter conducted his research on RNA polymerases and transcription of microorganisms with Zillig at Max Planck, Kandler continued working on the phylogeny of bacteria based on the composition of cell walls. Kandler was particularly interested in the cell walls of methane-producing "bacteria" and had sent one of his students, Josef Winter, to the United States to work in the laboratory of Ralph Wolfe at the University of Illinois. Kandler was interested in the methanogens for a number of reasons, one being that he believed they might have practical use in bioremediation since the microbes were commonly found in sewage sludge. Wolfe, who also was interested in bioremediation, had become one of the world's leading experts at cultivating methanogens, having improved on Hungate's methods of growing the oxygen-sensitive organisms.[3] Wolfe's breakthrough was growing the methanogens in pressurized bottles that evacuated all oxygen, allowing the anaerobic microbes to thrive. Wolfe became a wizard at growing methanogens in massive quantities.

Working with a small sample of methanogens provided by a culture collection from a laboratory in Munich, Kandler became especially intrigued by the unusual composition of the cell walls. Unbeknownst to Kandler, Carl Woese also was puzzling over oddities in the methanogens.

"Wolfe became very skilled at Hungate's methods and improved them to grow very large quantities of methanogens in culture. Wolfe could grow a kilogram of methanogens," Stetter said with amazement. "Around this time, Kandler had recommended that Josef Winter study in Wolfe's lab in Illinois. Wolfe's main interest was in the biochemistry of methanogenesis."

No one at the time knew that methanogens existed in super-heated environments such as those at hydrothermal vents on the ocean floor. But methanogens were known to exist in thermal environments in sewage growing at temperatures up to 60 degrees C.

"The methanogens were very interesting and known to make biogas. But they were taken as strange microbes, and very few microbiologists were able to grow these in pure culture. The most prominent was Wolfe, who was using the pressurized bottles to grow methanogens in pure culture. It was a very special accomplishment to grow methanogens in pure culture. This of course was where Carl Woese was working and developing his concept of the molecular taxonomy," Stetter explained. "Then Ralph Wolfe gave Carl Woese some of these methanogens he had cultured from the sewage sludge to look at, and this was amazing luck for Carl and the rest of the world of being in the right place at the right time."

He said, "In Germany, Otto Kandler was always wanting to look at the cell wall compositions of as many organisms as he could get to study their phylogenetic relationships. Someone at the German culture collection had grown one methanogen for Otto. He had identified a very strange cell wall already at this time and was suspecting something strange about the methanogens. Otto published this in a scientific journal. Since Otto knew Wolfe had such a great collection of cell masses he asked Wolfe to please send some cell mass, some freeze-dried methanogens, to his lab in Munich. Otto now was looking also at the cell walls in addition to the chemical composition, and he had a routine procedure for looking at these features of the cells. When he received his cell mass and looked at the cell walls of the methanogens he was very surprised. Kandler said this is very strange because he could not find murein."

Murein is a kind of peptidoglycan. It's made up of homogenous chains of sugars and amino acids that form a layer outside the cell membranes of bacteria. Think of an M&M candy with its hard sugary shell as the cell membrane. The peptidoglycan would be the thin coating of color on the shell. This layer is the same in all bacteria except that it's relatively thick in gram-positive bacteria and much thinner in gram-negative bacteria. Peptidoglycan chains are the target of penicillin and similar classes of antibiotics.

"Otto could not detect one molecule of murein in any of these methanogens, but he detected a different kind of polymer that was another peptidoglycan, which Otto called 'pseudomurein.' It was very exciting and so strange to see these different cell walls. The methanogens only had the pseudomurein," Stetter said.

The time frame for this work was in the mid-1970s. In the summer of 1977, Kandler was invited to give a talk at a Gordon Research Conference in the United States, and while he was there he traveled to the University of Illinois to meet Wolfe and discuss his new findings. Kandler and Wolfe were involved in a collaboration on the methanogens with Wolfe focusing on methanogenesis and with Kandler interested in the structure and composition of the cell walls. Kandler was unaware of Woese's work and had not planned to meet Woese during the visit to Wolfe's lab. Woese had not yet published his work and had shown his studies only to a handful of puzzled colleagues.

"When Kandler went to Wolfe's lab, Kandler told Wolfe, 'These methanogens look very, very strange, not even like bacteria. I have looked at hundreds of bacterial cell walls, and these for me are not bacteria!' Wolfe was an important celebrity among microbiologists and courageous to explore methanogens, and he did great work. But, considering phylogeny, Wolfe was very critical like the others. So he told Kandler 'Go a few doors down. There is someone named Carl Woese who says the same thing as you.' So Kandler went to talk about this with Woese and heard his concept of this new domain of life, which Woese called 'archaebacteria,'" Stetter said, becoming more animated and gesturing with his knife and fork. "Then as the first rumors started to make their way about this new concept and before the publication of the paper, Kandler told me about it. But Carl Woese is a little bit shy, and some people considered him a little bit strange. Woese had his own scientific way." Stetter wagged his knife and chewed.

Kandler and Woese became very close scientific colleagues and friends as a result of the meeting. For his taxonomic work, Woese was looking at only a small percentage of base pairs of sequence of one gene, the 16s rRNA. The scientific community was asking how it could be possible to determine taxonomy of a microorganism from just a few bases of sequence. But Woese knew the segment of coding that he was looking at in hundreds of species of bacteria was highly conserved and did not change. When Woese found that the sequence was different in the methanogen Wolfe provided, he knew something was up.

"Ralph Wolfe had been a very influential personality and was a member of the National Academy of Sciences. He also thought Carl had discovered something brand new. So Otto came back, and he said he was fully convinced of this new concept of Carl's. For Carl, this was the first time somebody outside of his own lab really recognized the importance of this discovery."

Stetter leaned forward and lowered his voice for a moment. He talked about the ridicule that several key American scientists laid at Woese's doorstep. He straightened back up and slapped his hand against the table. "But in Germany, Kandler said 'I'm convinced this is true, and let's explore the archaea continent!' Then I and Zillig said 'Let's look at the polymerases!' "

"Wolfram Zillig taught me molecular biology and biochemistry. And this was a wonderful time when we worked together on the first archaea. This *E. coli* person, Zillig, became excited about culturing the high temperature archaea," Stetter said, talking about the strange microbes Brock discovered. Thomas Brock studied microorganisms in the hot springs of Yellowstone National Park throughout the 1960s and 1970s. Brock published extensively on sulfur-loving microbes as well as salt-loving microbes from brines and salt lakes. Brock also discovered microbes that thrived in burning piles of coal refuse. Stetter's curiosity led him to attempt to culture the odd microbes.

In 1978, Josef Winter returned to Germany from Wolfe's laboratory, bringing with him the knowledge he gained for cultivating methanogens. Winter taught Stetter how to grow the methanogens and how to purify oxygen-sensitive enzymes—the RNA polymerase. Stetter said he tried to purify the enzymes from methanogens on his own, but he had problems keeping the microbes from being destroyed by oxygen. In need of a larger amount of methanogens, he contacted Wolfe and asked for more.

"He sent me a big quantity of frozen methanogens. This was a real treasure. I tried to purify the DNA polymerase enzyme, and when I had succeeded it looked very different from other polymerases," Stetter said.

Stetter continued the research on transcription in different microbes. He learned from Zillig how to use an antibiotic called "rifampicin" to block transcription in bacteria. All of the various species of bacteria studied by scientists over the years were sensitive to the antibiotic. Exposing microorganisms to it became a standard method for discerning whether an unknown species of microbe was bacteria or a eukaryote. Eukaryote cells are resistant to rifampicin. Stetter exposed the methanogens to rifampicin to see what would happen. If methanogens were bacteria their transcription would be disrupted. If they were something else, the transcription would be normal.

"My experiments with methanogens from the cells of Wolfe came back with the cultures completely resistant. We thought all bacteria were sensitive to rifampicin. As it turns out the archaea have a primitive transcription system the same as the eukaryotes. But of course they are not eukaryotes because they do not have a cell nucleus. So, the methanogens were the first archaea to be found," Stetter said.

Meanwhile, in the mid-1970s, another group of scientists conducted research on lipids in the cell membranes of the entire ensemble of microbiology's freak circus—the salt-loving halophiles, the sewage sludge methanogens, and the sulfur-loving thermal microbes from Yellowstone. Morris Kates of the University of Ottawa led this team. Kates, who is now retired, was considered the leading scientist in the field of lipid research. His textbook *Techniques of Lipidology: Isolation, Analysis, and Identification of Lipids* is a classic. Kates is the person who discovered that the bonds and chain structure of lipid molecules in the cell membranes of the extreme microbes are different from those in all other bacteria. The molecular bonds in archaea are much stronger and structured in branches instead of chains. Unlike bonds in bacterial cell walls, they cannot be easily broken by hydrogen. Since the halophiles, stink bugs, and methanogens all use hydrogen as energy, the differences found between the bonds made sense on the surface. Again, the weird membrane lipids were thought to be independent adaptations by organisms to their extreme environments. Kates had not yet heard of Woese's work so convergent evolution was the best possible explanation within the two-domain paradigm of prokaryotes and eukaryotes.

Kates told *Phoenix* magazine in a 2002 interview:

> The highlight of my scientific career, from a professional point of view, was the discovery that membrane lipids of halophiles (bacteria living in environments of high salt concentration), methanogens (anaerobic autotrophs that obtain energy from the synthesis of methane gas), and thermo-acidophiles (bacteria growing at high temperature and low pH) have structures different from all other organisms and are synthesized through a very unusual and unexpected pathway. Our findings provided an important clue that was used by Carl Woese in proposing the existence of a third class of organisms called archaea, to which the extreme halophiles and methanogens belong. A survey of lipids from many species of extreme halophiles showed there was a good correlation between the lipid structures and the genus of the species and made it possible to classify these species taxonomically on a generic level.

Stetter and Zillig's work on the polymerases would prove vital to supporting Woese's theory of a third branch on the tree of life. When Woese announced the existence of archaea in 1977, he had only the cell membrane data from Kandler and his own 16s rRNA sequences as proof. Other data existed, but Woese had not yet connected all the dots.

"There had been at this time a kind of patchwork phylogeny. Other people were looking at enzymes, and a third group was looking at lipids. But none of these guys thought this was important," Stetter said. "Brock thought the differences in the cellular features represented convergent evolution and that none of the organisms—the thermophiles and halophiles—had anything to do with the other. The overall view was completely missing."

Stetter and Zillig continued working on cultures of other candidate archaea, finding similar differences in the RNA polymerases, all in support of Woese's proposal. Stetter had no doubt that archaea represented a new domain of life.

"In 1980, after Woese was beaten up and nobody was believing in him, Zillig, Kandler, and I organized a meeting in Munich and invited Woese, and everybody was ringing a big bell for the archaea," Stetter said while gesturing with his fork like a conductor punctuating the final triumphant notes of a symphony.

Kandler expressed heartfelt sympathy for Woese's trials. In particular, Kandler was disappointed in the way such a prominent figure as Nobel Prize winner Salvador Luria treated Woese. It became known that Luria attempted to turn Wolfe against his shy and reclusive colleague. Kandler arranged for a band to play when Woese entered the meeting in Munich. Woese received the honor with great relief and became close friends both with Kandler and Zillig. Woese is quite fond of the picture taken in 1981 of Kandler, Wolfe, and himself at Rofan Mountain in Austria. Woese had longish hair and a beard, looking rather dashing and hip for a shy recluse. Woese visited Kandler's lab in Munich on several occasions. He saw firsthand the work Kandler was doing with peptidoglycan and visited with Zillig and Stetter at the Max Planck Institute where the duo were pursuing transcription of methanogens and other suspect species of "bacteria." German microbiology laid a solid foundation to become the world leader in archaea research.

"Woese's ideas were initially more accepted in Germany. In contrast, in the beginning several people in the United States did not accept Woese's theory, and some became very personal and wanted to make a fool of him. This made Carl bitter." Stetter lowered his voice and said with a hint of sadness, "Just very recently when Carl was given the Crafoord Prize he described how badly he had been treated in the front of the King and Queen of Sweden. He said to them that he discovered the archaea but nobody liked them. Of course this was true," Stetter said, leaning forward and looking straight over the top of his eyeglasses.

Most accounts of the history of archaea describe the discovery as one made by Woese and Wolfe. While Wolfe certainly suggested that Woese look at the methanogens to see where they fit on his tree, and privately became a staunch ally, Wolfe was publicly more cautious. Wolfe's name does not appear on the paper that proposed "archae-bacteria" as a third domain. Woese's co-author was his postdoctoral student George Fox.

The vast majority of the substantial foes of Woese's third domain never visited his lab or examined the significant evidence gathered by Kandler, Zillig, Stetter, Kates, Pace, and others who pursued Woese's theory of a third domain. They simply ignored it. Remember what Woese said at the beginning of his address to the Royal Swedish Academy of Sciences?

"The Archaea were unexpected to begin with, and having arrived, they were unwelcome. This particular point needs understanding in that the situation is not unique to the Archaea." And recall he also said: "So rather than question the dogma, most biologists were content to condemn the finding. Their failure to question that dogma is one of the black marks in microbiology's history."

Mathur compares Woese's trials to the way Galileo was treated for confirming and publishing work that supported Copernicus's revolutionary hypothesis that the Earth moved around the sun. Stetter says it was shocking to see how Woese was treated by the scientific community in the United States.

"People were saying this archaea business would disappear in a few years, but they were wrong!" Stetter said with absolute satisfaction. He took his napkin, dabbed his mouth, and relaxed back in his chair. The steaks—our hydrogen—were consumed and already being digested by enzymes in our stomachs, fermented by lactobacilli in our intestines, and converted through our aerobic metabolisms into fuel. We ordered coffee and dessert.

"With Carl Woese's idea and our own data this changed everything for Wolfram and myself. Now we were looking hard at these tough guys, these microbes that lived in the crazy places. First there was halobacterium, then came next the news of *Sulfolobus.* For my research on the archaea I was going back and forth between the Botanical Institute at the University of Munich and the Max Planck Institute. I had two labs."

From 1977 to 1980, Stetter and Zillig continued their work on RNA polymerases with an increasing number of new organisms. They worked with methanogens, halophiles, sulfur-loving thermophiles, and finally an odd microbe known as *Sulfolobus acidocaldarius.* This organism Brock had isolated from a Yellowstone hot spring in 1972 and misclassified as bacterium.

"We saw that this organism grows at 75 degrees C. I would later identify these organisms as hyperthermophiles. Wolfram was interested in learning how to grow *Sulfolobus* on a large scale. We began to purify the enzyme and again, like a eukaryote, it was resistant to rifampicin. Then came news from Carl Woese based on 16s rRNA sequence that Brock's *Thermoplasma acidophilum* belonged to archaea. We also wanted to culture this and purify its enzymes, but this was not at all easy to get growing. It grows naturally in smoldering coal refuse piles, and it is growing naturally at a very acidic pH."

Brock and his team at Indiana University in Bloomington in the 1960s isolated *Thermoplasma acidophilum* from smoldering coal at a mine in Indiana. Brock published a paper on it in 1970.

Since publication of Woese's paper in 1977, Stetter had been working like a detective tracking paper trails in the form of scientific publications on extreme forms of bacteria. Most of these trails led back to Brock, who is still a legend at Yellowstone National Park. Brock wrote a guide to the microbes of Yellowstone for the public and served as the key inspiration to Stetter to search for his own microorganisms.

"Tom Brock never found *Thermoplasma acidophilum* in Yellowstone hot springs, but we found them in hot springs growing up to about 60 degrees C, very acidic and with no cell wall at all. It is just surrounded by cell membrane. This organism also was resistant to the rifampicin," Stetter said.

By 1980, Stetter began to thirst for more species of archaea, especially ones that might exist at extremely high temperatures. It was at this time that Stetter and Zillig made a trip to Iceland, known for its volcanic hot spots.

"It had been with Wolfram Zillig that I had my first sampling trip to Iceland in 1980 to search for *Sulfolobus* and to look for new high temperature organisms there. We almost killed ourselves when we were sampling a couple of times," Stetter said. "You have to be very careful not to fall in the hot pools or break through the thin crusts that surround the hot springs. You can break through the surface, and then you are in boiling water."

Stetter told another story of having a picnic with his wife on a flat rock one day and returning later to find a massive geyser erupting in the same spot. Stetter laughed.

"As you know, every living being is adapted to a specific growth temperature. In the case of men, this is 37 degrees C and already an increase by 5 degrees C becomes fatal. In the microbial world, the temperature range of growth is much more diverse. As I have been preparing my speech for the Leeuwenhoek Medal, I am wondering why no other groups before us had isolated microbes thriving in boiling water."

The Leeuwenhoek Medal is given once every 10 years for the most significant achievements of microbiology in the previous decade. Stetter received the medal five weeks after our trip. The previous medal, awarded in 1992, went to Woese. Stetter explained:

> I think it was not a matter of laboratory equipment. In 1979, several groups had much better cultivation facilities than I. Tom Brock had already reported on non-culturable microbes in boiling hot springs. However, to my feeling it was because of the more common view of thermophiles at that time that prevented their isolation. Tom Brock's Sulfolobus, which grew already at temperatures of up to 85 degrees C, had been considered to be a kind of very rare curiosity. The idea was that various recent evolutionary "tricks" in biochemistry may have led to its unique heat adaptation. Its aerobic lifestyle appeared essential to having made this adaptation possible. The much lower growth temperatures of anaerobic thermophilic methanogens like those Wolfe provided to us seemed to confirm this kind of view. No one believed that there are hyperthermophilic methanogens. High salt concentrations were taken as additional adaptive stress for the hot lifestyle. Therefore, at that time the possibility of life in boiling water within anaerobic and marine environments had never been taken into consideration.

In essence, no one looked for microorganisms in boiling water because no one believed they could exist there. But Stetter and Zillig were determined to find new organisms in the hottest springs possible and be the first to report on them. They wanted to discover more anaerobic organisms other than methanogens, but they had to learn how to find them. Temperature is easily measured by inserting a thermometer attached to the end of a pole. But how does one go about determining that a pool of water contains no oxygen? Stetter and Zillig used a dye called redox resazurin to inject into hot springs. The dye is blue. If there is no oxygen in the hot spring, the resazurin is "reduced" and becomes colorless. If oxygen is present a blue cloud spreads into the water. Nobody had ever done this before.

"I injected clouds of resazurin into several hot springs in Iceland, which became immediately reduced, indicating that these environments were strictly anaerobic," Stetter said, promising to demonstrate this on a sampling trip in the future. "From a sample taken from a sulfataric field in the mountains in Iceland, I isolated *Methanothermus fervidus*, which was a brand new methanogen, and I discovered in my laboratory that it grew fastest at 82 degrees C and exhibited a maximal temperature of growth of 97 degrees C. That meant we had discovered a strictly anaerobic organism that grew at much higher temperatures than the aerobic *Sulfolobus*! The sulphur-respiring *Thermoproteales*, which had been discovered together with Wolfram Zillig at the same trip, exhibited similarly high growth temperatures and were again anaerobes."

This was the beginning of many adventures into the outer limits, and it launched a series of dozens of papers that Stetter would publish with Zillig and later on with other colleagues at Regensburg.

Mathur generously picked up the considerable tab for dinner. The dining room at Chico Hot Springs is popular with the Hollywood crowd that hangs out in Montana. The prices were accordingly adjusted upward. We walked back to the porch of Stetter's cabin to have Cuban cigars. The Montana night sky blazed with stars. I told Stetter about meeting Carl Sagan in 1992 and his conviction that microbial life was flourishing throughout the universe.

"Of course," Stetter said. "Why not?"

Mathur nodded and released a cloud of bluish smoke that drifted upward into the air and blended perfectly with the Milky Way. "Just wait, man. Wait 'til you hear more about the *Nanoarchaeum*. They look evolved for interstellar space travel—maybe the most opportunistic life form ever discovered."

The next morning after breakfast we drove on to Yellowstone. Stetter handed me a draft of his paper on the sequence of *Nanoarchaeum*, which I attempted to read on the two-and-a-half hour road trip. But I'd never been to Yellowstone and couldn't keep my eyes off of the scenery. The first Thermal Biology Institute conference was held at the Old Faithful Snow Lodge, which was the only lodging still open in the park in October. The original Old Faithful Lodge, a massive log structure built in 1928, was closed for winter except for the lobby and restaurant. The day's plan was to stop about halfway to the lodge at Norris Geyser Basin. I didn't understand the significance of why we were stopping there until we arrived. The site was closed at the time to the public except for the bookstore and its little museum run by a single park ranger. Norris Geyser Basin had become geothermally unstable over the past few months. The area has the hottest springs in Yellowstone, which of course was why Stetter wanted to stop. He sampled there numerous times with his bamboo pole and stainless steel cup.

In 1929, scientists from the Carnegie Institute of Washington, DC, drilled test wells to determine the temperatures of the underground springs. One of the wells had to be abandoned at 265 feet when the temperature reached 401 degrees F and steam pressure threatened to destroy the drilling rig. After lying dormant for about a decade, Steamboat Geyser in the Norris Basin began erupting in 2000. It erupted twice in the spring of 2003. We arrived midday on October 8. It erupted again October 22 in the same spot Stetter led us to after showing the park ranger his scientific credentials. The ranger cautioned us to be careful and looked at us like we were crazy.

We began hiking down a hill to a boarded walkway. Temperatures of 200 degrees F had been measured in several places here less than a centimeter beneath the soil and crusty mud. New springs and mud pots had been appearing. Stetter found a recently dead tree branch about six feet long on the ground and picked it up as we made our way downward. As we reached the basin, the trees and plants were all dead and dying due to the high ground temperatures and acid levels rising in the soil since July.

Wisps of steam swirled above the surfaces of the hot springs and mud pits, all bubbling like cauldrons in a scene stolen from Macbeth. The slight breeze carried the primeval scent of hydrogen sulfide. The water in the springs was about 198 degrees F with a pH of 2. Stetter cautioned us not to go swimming with the stink bugs and laughed. This actually happens to an unfortunate tourist, park worker, or pet every few years. We were finally in the outer limits: archaea's North American playground.

Nothing like a little deadly fire and brimstone to achieve the perfect setting for the fearless Karl Stetter, microbe hunter. If Carl Woese can be considered the King of the Archaea Domain, then Stetter is its Crown Prince of Darkness. He stepped from the safety of the boardwalk that skirts an area called Porcelain Terrace Overlook. This is forbidden, of course. But Stetter is accustomed to hellish places; loves them. He used the long stick to poke at the white- and yellow-crusted ground before him. Cautiously, he stepped toward the edge of a spring. The end of the stick broke through the crust releasing a puff of steam and stinking gas. "Aha! You see how dangerous this is!" Stetter exclaimed. He muttered to himself, "I should have my rubber boots."

Stetter waved his staff at the bubbling springs as clouds of hydrogen sulfide curled upward and danced at his feet. Prodding at the brimstone, Stetter conjured images of Gandalf. His dynamic personality and selfless immersion into the peculiar habits of archaea have made him perhaps the most popular lecturer in the growing field of archaea-obsessed microbiologists. Stetter's infectious laugh and devil-may-care spirit are not commonly observed in American scientists. It is easy to imagine Stetter at age seven filling jars of pond water to view under his new microscope.

Stetter told the tragic story of a young tourist who dove into a hot spring here on July 20, 1981, to save his dog. The tale is recounted in *Death in Yellowstone: Accidents and Foolhardiness in the First National Park* by Lee H. Whittlesey. According to Whittlesey's gruesome account, David Allen Kirwan and his friend Ronald Ratliff walked down to the hot springs near the parking lot of Fountain Paint Pot. They left Ratliff's dog, a great dane or mastiff, in their truck. The dog escaped from the vehicle and followed the two young men. Upon seeing the water, the dog bounded into Celestine Pool. Kirwan began undressing to rescue the dog as people shouted warnings to him not to get into the water, Whittlesey wrote. The temperature of Celestine Pool is 202 degrees F. Kirwan ignored the warnings and dove headfirst into the water to save the yelping dog. The pH of the spring is around 9.

One witness saw "Kirwan actually swim to the dog and attempt to take it to shore, go completely underwater again, then release the dog, and begin trying to climb out. Ronald Ratliff pulled Kirwan from the spring, sustaining second degree burns to his feet." The witness saw "Kirwan appear to stagger backwards, so the visitor hastened to him. . . ." The visitor walked with Kirwan and "was suddenly overwhelmed with the feeling he was walking with a corpse. He could see that Kirwan's entire body was badly burned as the skin was already peeling off." Kirwan's eyes had turned completely white. As another witness attempted to remove one of Kirwan's shoes, the flesh of his foot pulled off too. "Near the spring, rangers found two large pieces of skin shaped like human hands," according to Whittlesey. The dog was left in the spring. Oils from the dog's body caused little eruptions in the spring. Kirwan died the next morning at a hospital in Salt Lake City.

Whittlesey's book has several horrid accounts of people falling into Yellowstone's hot springs dating back to 1871. Deaths from hot springs are more common than deaths by grizzly bears. The white bones of many animals, including elk and buffalo, have been found in the hot springs. At least nine children have perished there. Summer workers also occasionally meet horrible deaths. It is easy to see why we assumed life cannot exist in such hellish springs.

Stetter found solid ground near one of the springs and summoned us from the boardwalk for a look. The water was bubbling and crystal clear. The colors of the rock in the pool change with gradients of temperature, from orange at the center of the spring where the water is hottest to increasingly lighter shades of yellow and then green at the outer rim. Each color represents different species of microbes, from anaerobic archaea at the heart of the spring to oxygen-using cyanobacteria at the edges. Pink *Aquificales* swirl in the current downstream. The color gradients of a two-foot-diameter pool represent the whole of evolution from the first anaerobic microbes to photosynthetic oxygen-lovers.

The afternoon sky darkened, so we decided to walk back to the car. Without the sun, the temperature drops suddenly and the wind picks up. After the tales of unfortunate deaths, and the recent geothermal instability, the Norris Basin felt creepy and dangerous. No one spoke much after departing for the Snow Lodge and the Thermal Biology Institute conference, sponsored by Montana State University (MSU). MSU became a Mecca for scientists with archaea fever, able to work full-time in close proximity to the thousands of hot springs at Yellowstone. After Norm Pace's lab, it's the "It location" for extremophile research in the United States.

Stetter's star status was immediately apparent when we arrived at the lodge. Professors and students milled about the lobby wearing plastic name badges. This was a small meeting with about 50 attendees. As Stetter checked in, several young scientists, Ph.D. candidates and postdocs, approached from behind and hovered. Sensing that he was being staked out, Stetter turned around and greeted them. One young woman actually blushed as Stetter shook her hand. A young man in her group pulled out a digital camera and snapped a picture certain to be posted on a lab wall or captured as a screen saver. Stetter agreed to meet them in the bar after he settled in. Stetter, too, was anxious to speak with the scientists here. They are hard-core archaea-philes. In the aluminum briefcase Stetter was carrying were the new data he presented on *Nanoarchaeum.*

My immediate mission was tracking down John Varley, the scientific research director for Yellowstone National Park, to get the broader picture of research at Yellowstone and discuss the biodiversity agreement he worked out with Diversa in 1997. I found Varley talking with friends in the lobby. He has a bushy mustache and resembles Teddy Roosevelt, only better looking. He wore khakis and a park-ranger green dress shirt. One gets the impression he has his hands full managing bioprospectors and fighting federal bureaucracies, and that he absolutely loves microbes. In addition to wolves, grizzlies, bison, and all the other wildlife research, Varley oversees all studies conducted by academics and private companies sampling microbes in Yellowstone's hot springs. He gave a talk titled "Biodiversity and Bioprospecting in the World's Natural Icons: Can We Keep Both Greens and Biotechnology from Spoiling a Good Idea?" The title suggested one source of his frustration. The Greens—the anti-biotech, environmental activists—have been suing Yellowstone for five years to stop the royalty-sharing agreement with Diversa. The lawsuit does not prevent companies from working in the park. If successful the lawsuit would mean that Yellowstone can't earn royalties if any of them strike microbial gold. The fact that the park lost out on *Thermus aquaticus* exploited for PCR back in the 1980s is something foremost in Varley's mind. It is only a matter of time before another biotech prospector makes another big strike.

The benefits-sharing agreement worked out between Yellowstone and Diversa, which includes financial royalty and data sharing, became Yellowstone's model for all bioprospecting endeavors in the park's hot springs, including those being done by academics. In today's world, every university has a technology transfer office ready to patent anything discovered by academics that might make money in the future, and most university researchers involved with genes of any kind have a plan for a biotechnology company in their back pockets.

Varley and I found a quiet spot in the rustic lobby and sat down. "The idea of setting aside some of the most beautiful and remarkable places in the world as national parks began with Yellowstone 131 years ago and blossomed into over 10,000 parks and equivalent reserves in over 160 nations," Varley said. "These pristine wild lands harbor a tremendous biological diversity that attracts researchers and bioprospectors. Managing human activities in parks while preserving the natural values they were established for is a challenge."

There's an understatement. Everyone who sticks a sampling pole in a Yellowstone hot spring is supposed to come to Varley first. The number of people seeking permits has tripled in just a few years. Yellowstone has a long history of microbial research. The first known research permit to collect microorganisms was issued in 1898 at Mammoth Hot Springs. W. A. Setchell of the University of California believed he was studying heat-loving algae colonies. Setchell wrote: "All of the strictly thermal organisms are low forms, not even representing the high differentiation in the group in which they belong." Setchell may have been examining the pink streamers of bacterial *Aquificales.*

Varley oversees permits issued to more than 50 microbial research projects at Yellowstone. In the prime summer sampling months, Varley may have more than 100 scientists at any given time trekking through the park and dipping their poles. NASA is among the groups holding research permits. NASA considers Yellowstone to be "one of the world's best-preserved windows to understanding the origin of life on Earth as well as the possibility of life elsewhere in the universe," Varley said proudly.

Yellowstone has an estimated 10,000 thermal features, including hot springs, fumaroles, boiling mud pits, geysers, and hydrothermal vents. "Modern molecular biodiversity prospecting, now about 30 years old, is the newest unregulated human activity and it has been vexing to resource managers because of its complexity," Varley said. "The Yellowstone model of managing bioprospecting and benefits-sharing differs from other accepted international models in important ways, but it's consistent with the concept and spirit of the Convention on Biodiversity. Our model has broad acceptance in United States industry, university, and government institutions, and it's been largely apolitical. But some environmental activists and a few scientists vigorously oppose it in both concept and practice."

The bioprospecting agreement Varley developed with Jay Short allows companies to conduct research on microorganisms discovered in the park in exchange for royalties on future profits from any products developed from the research. Yellowstone can apply the royalties to conservation programs and scientific and public education activities.

The main difference between the Yellowstone model and those in other countries is Yellowstone also receives scientific data generated from a company's research in the park. Yellowstone uses the data to bolster its own biodiversity management program.

"The agreement with Diversa has generated interest within the wider conservation community, research institutions, and scientific and educational organizations that want to know how research-focused bioprospecting can benefit resource conservation activities," Varley said. "Yellowstone is the first U.S. national park to develop benefit-sharing arrangements designed to conserve the biodiversity it protects and manages."

The World Foundation for Environment and Development assisted Varley and Diversa in their negotiations. Since 1997, the agreement has become the model for agreements developed with about a dozen other biotechnology companies through 2005. From Varley's perspective, obtaining scientific data from companies and academic researchers is as important as royalties. "Through properly structured agreements with the research community, Yellowstone can obtain improved data relating to microbial distributions throughout our various thermal systems from visiting researchers. We can do a better job of protecting our resources by directing scientists to thermal areas known to provide habitats for desired organisms found in less-frequented areas of the park."

The application of *Thermus aquaticus* to PCR launched the microbial gold rush. Varley mentioned the $300 million sale of Cetus's PCR patents to Roche Molecular Systems and shook his head. "Lacking an arrangement like the one we have now, Yellowstone has not shared in these revenues. Yellowstone's protected areas represent the greatest concentration of thermal habitats on the Earth's surface. The abundant concentration of thermophilic microbes here provides an important opportunity to explore new ways to support microbial research as well as to improve public understanding about these little-known resources. With the heightened scientific interest in Yellowstone's microbial diversity we can enhance the prestige of the role of national parks *and* microbes in American life."

Varley has been directing key research at Yellowstone for two decades. But his greatest legacy for the park may be the biodiversity agreement. The next time a company hits the jackpot with Yellowstone microbes, which is certain to happen, the park's scientists will get new labs.

A small group of scientists gathered about 10 feet away and waited politely for Varley to wrap up so they could grab a beer and talk shop. I turned off the tape recorder and closed my notebook to everyone's relief. As they made their escape I took a closer look at the program. The bell ringer at 8 p.m. was subtly titled: "Volcanic and Hydrothermal Processes Above a Large Magma Chamber: Results from High Resolution Sonar Imaging, Seismic Reflection, and Submersible Surveys of Yellowstone Lake."

The key phrase here is "a large magma chamber." A massive swath of Yellowstone National Park is a caldera—the crater of a supervolcano that has erupted like clockwork over geological time. Much of Yellowstone Lake sits over this caldera. The supervolcano erupted 2 million years ago, 1.2 million years ago, and again 600,000 years ago. Scientists gauge the last one as about 1,000 times more powerful than the eruption of Mount St. Helens. Since about 1980, geothermal activity beneath Yellowstone has been increasing. This is why Norris Geyser Basin was closed to visitors. If another eruption occurs similar to the one 600,000 years ago, say goodbye to the Pacific Northwest and to life as we know it in that region. Archaea's playground would expand dramatically.

"I don't think visitors appreciate that they're standing directly on top of the largest, most dynamic magmatic system on the planet," geologist Daniel Dzurisin told science writer Jessica Marshall in the March 1, 2006, *NewScientist.* Dzurisin, Charles Wicks, and a team from the U.S. Geological Survey made measurements from radar images of the caldera taken by the European Space Agency's ERS-2 satellite in 2005. Their results published in *Nature* showed that a circular region of the northwestern rim of the caldera is rising as southern portions of the caldera sink. Thousands of acres of Earth's surface don't normally tilt in a relatively short time span without a significant reason.

Initially, scientists believed the uplifting, which created an ominous dome on the bottom of Yellowstone Lake, was due to hydrothermal gases. That wouldn't be so bad. Gases can vent through cracks on the bottom of the lake and through new geysers and through the 10,000 or so other features at Yellowstone. The worst that could happen from a dramatic eruption of gases might be a bubble that empties half of the lake and maybe creates a big fireball that rains down and reduces Yellowstone to cinders. Yellowstone Lake has numerous hydrothermal vents on the bottom but the extent of the vent system was unknown until now. Next to Lake Baikal in Russia, Yellowstone is the largest site of freshwater hydrothermal vents in the world.

As it turns out, the source of the surprise rising of the northwestern rim appears to be the movement of magma seven miles below Yellowstone. This is not good. A catastrophic eruption could occur in our lifetime or maybe not for another 100,000 years. If Yellowstone's supervolcano erupts again, computer models suggest it would create a cloud of ash 4,000 miles wide in one day. Hello nuclear winter and the rise of the second Archean Empire.

Lisa Morgan of the U.S. Geological Survey, who delivered the first evening's presentation, showed an incredible video made by a remotely operated vehicle (ROV). Among the most striking findings made by her team was the discovery of more than 300 new hydrothermal vents on the bottom of Yellowstone Lake.[4,5] They also found several old hydrothermal explosion craters more than 1,600 feet in diameter—of the type that would make a big bubble—and new domes that could become future explosion craters. The ROV also highlighted a mind-blowing grouping of siliceous hydrothermal spire structures up to 20 feet tall that kind of resemble big black and white smokers found along exotic hydrothermal vents on the ocean floor. But no one is certain what created these spires. They have internal pipes similar to smokers but the chimneys do not go through to the tops of the spires. The giant structures might have a microbial origin. More on this later.

As Morgan continued her presentation, Mathur leaned over and whispered, "We absolutely have to dive on these vents and the spires and take samples." I glanced over at him. He was dead serious. Neither of us knew it at that moment, but less than eight months later we would be back with scuba gear and sampling equipment, and slipping into the 38 degrees F water of the lake. I'm certain that Mathur went to bed that night plotting how to make it happen. I went to bed exhausted but thrilled to be there.

The next morning's presentations began at 8 a.m. with overviews of the hydrothermal systems of Yellowstone, a chemistry lesson on the hot springs, and a presentation on how organisms in hot springs generate energy from chemicals and heat. At 10 a.m., Karl Stetter took the stage, as the most breathless and dramatic lecturer I have witnessed.

The title of his morning talk was "A Novel Kingdom of Parasitic Archaea." He described the discovery of *Nanoarchaeum equitans* at the hydrothermal vent north of Iceland. The area is known as the

Kolbeinsey Ridge. Using the two-person research submersible Geo, samples were taken of sandy sediments and vent fluids at temperatures around 90 degrees C. Black smoker samples obtained during a dive made on the submersible *Alvin* at a vent in the Pacific also were analyzed. Initially, the samples from the mid-Atlantic ridge revealed a new genus and species of archaea, which Stetter named *Ignicoccus islandicus.* Electron microscopy photos taken at Stetter's lab of an additional *Ignicoccus* isolate revealed tiny strange spheres attached to its surface. This was shocking. No such thing had been seen on archaea. By culturing the organisms together Stetter was able to isolate *Nanoarchaea* then look for segments of its RNA. It does not possess the similar ribosomal RNA signature of other archaea.

"With a cell diameter of only 400 nanometers, it is one of the smallest living organisms known. Cells grow attached to the surface of a specific host, a new member of the genus *Ignicoccus hospitalis.* Owing to their unusual 16s rRNA sequence, members of *Nanoarchaeum equitans* remained undetectable by commonly used ecological studies based on the polymerase chain reaction," Stetter lectured. He was gesturing, pointing to features on his slides with sweeping motions of his whole arm. The organism has only 563 genes and has the most compact genome discovered so far—meaning it has very little if any "junk" DNA, which is plentiful in most multi-celled eukaryotes. The DNA content of *Nanoarchaeum* is close to the amount predicted to be minimally essential for life, he says. The presence of *Nanoarchaeum* on the surface of *Ignicoccus* suggested that it was either a parasite or a symbiont. And it appears to be extremely ancient. It remains dormant when it is not attached to its host.

"After sequencing of the total genome, the phylogenetic relationships of *Nanoarchaeum equitans* could be investigated," he said.

Analysis of the amino acid sequences of 35 ribosomal proteins revealed that *Nanoarchaeum* was almost as different from other archaea as archaea are from bacteria and eukaryotes. To Stetter this meant that *Nanoarchaeum* belonged in its own branch within the archaea domain. At the time, scientists determined that the known species of archaea should be divided into three branches. Stetter was announcing that *Nanoarchaea* was a fourth! The sequencing of *Nanoarchaeum* was a collaboration among Diversa, Celera, and the Department of Energy's Joint Genome Institute (JGI). Today, the number of archaeal branches is in dispute. Norm Pace's group suggests only two super-branches of archaea exist. Pace argues that *Nanoarchaeum* is a sub-branch of Euryarchaeota. Regardless, Stetter says *Nanoarchaeum* is of a deeply branching, very ancient lineage.

"*Nanoarchaeum* was placed with high support at the most deeply branching position within the Archaea, suggesting that the *Nanoarchaeota* diverged early within the Archaea. With only 490,885 base pairs, *Nanoarchaeum equitans*'s genome is the smallest microbial genome sequenced to date and also the most compact," Stetter said.

Sequencing revealed that 95% of *Nanoarchaeum*'s DNA likely represents coding regions for proteins or RNA. What is most incredible is that the microbe has all the genetic machinery needed for information processing and DNA repair, but lacks genes for lipids, amino acids, cofactors, and nucleotide biosynthesis. It needs to borrow another organism's metabolic engine. *Nanoarchaeum* might be able to manipulate the metabolisms of just about any archaeal host. Lacking genes for lipids make one wonder whether it also can use single-cell eukaryotes or bacteria as hosts. What versatility that would be. Since Stetter's presentation, *Nanoarchaeum* has been discovered on other species of archaea from the Uzon Caldera in Kamchatka, China, and Yellowstone.

"Two 16s rRNA sequences from Uzon Caldera and Yellowstone National Park exhibited 83% sequence similarity to *Nanoarchaeum equitans*. Light microscopy and fluorescence *in situ* staining reveal that these are new *Nanoarchaeota*," Stetter said. Nanoarchaeota is the name he chose for the controversial new branch. The new specimens are about the size of *Nanoarchaeum equitans*, but attached to different hyperthermophile hosts.

"The *Nanoarchaeota* appear to be globally distributed and had been completely overlooked so far. The discovery suggests that further different groups of microbes may be undetectable by our present knowledge and are waiting for us to find them in order to tell us more about the origins and evolution of life," Stetter said. He wasn't revealing it at the time, but Stetter was already working on this.

The *Nanoarchaeum* sequencing paper, published after the conference in the *Proceedings of the National Academy of Sciences* (*PNAS*), concluded, "The complexity of its information processing systems and the simplicity of its metabolic apparatus suggests an unanticipated world of organisms to be discovered." Indeed.

Follow-up studies conducted since the sequencing of *Nanoarchaeum* reveal that it has incorporated copies of some of *Ignicoccus*'s genes into its own genome. How or when this occurred is unclear, but JGI scientists suspect it occurred long ago in the history of the relationship between the two microbes. Stetter suggested *Nanoarchaeum* or something like it might be a predecessor of the nucleus of cells in the eukaryote domain. Who knows? It is certainly one of the oldest and perhaps the oddest forms of life known on our planet. As a parasite of other archaea, it could have an incredible survival advantage. It may be able to thrive in any environment with whatever type of metabolism a host happens to make available. I asked Stetter if *Nanoarchaea* could survive in space, waiting with its unique set of genes to find an archaeal host with any metabolism. He said *Nanoarchaea* do not create spores. But his eyes lit up when he thought of their possible survivability in the cold of interplanetary space, he said, "Why not?"

After the conference, sitting in an executive lounge at Salt Lake City's airport, Stetter and I used the time to discuss how he grows archaea and determines which nutrients they need. The key is mimicking as closely as possible the exact chemistries and conditions of the microbes' natural environments. Scientists are learning this also includes culturing groups of microbes together because they often rely on each other's metabolites. Early on, Stetter checked with the manager of a large aquarium for the exact formula for seawater and added whatever minerals were needed at the right concentrations. But it was an art to get the right mix, temperatures, and pressures, not too different from growing orchids. Whenever one of the world's space agencies gets around to bringing back a sample of soil from Mars, Stetter would most likely be the person to succeed at culturing alien microbes, if they are present.

Mathur told me over dinner one evening at the conference that when Stetter was learning to brew archaea a pressurized bottle once exploded in his laboratory. Another mishap, no fault of Stetter's on either occasion, resulted in hydrogen sulfide gas being vented into the building that housed his laboratory. I wanted to ask Stetter about the incidents, but wasn't sure how to phrase the delicate question. Mathur finally holstered his Blackberry and leaned forward. "You should see Karl's lab, man. It's totally out of this world."

Stetter glanced at Eric, thinking and nodding his head.

"Eric is right. You must come to Regensburg. I will show you where we grow the archaea and how it is that we do it."

I left for Munich two weeks before Christmas. Stetter picked me up at a hotel on the outskirts of Munich on a Sunday afternoon. Snow was falling. Stetter talked about retiring from the university on the snowy drive to Regensburg. He had recently turned 62. The German academic system requires professors to step down from their positions and make way for younger minds. Stetter was absolutely frustrated. He was still in the prime of his career and on the verge of new discoveries. Stetter was continuing to detect the undetectable, and culturing what no one knew how to grow. He would keep his laboratory and office for a time even though he was losing control over the University of Regensburg's Archaea Research Center, which he founded and had directed since the early 1980s. He maintained a significant laboratory at Diversa in San Diego where his Ph.D. student Jim Elkens was working on a project related to the isolation and characterization of the first Korarchaeota. It was too new to talk about it in any detail, but he could say that it was possibly groundbreaking.

By the time we reached Stetter's home in Regensburg the snow was six inches deep and still falling. His house was old, three stories and built from sturdy stones. It was surrounded by an iron gate on a wide tree-lined street. Stetter's wife Heidi greeted us at the door. Inside the stately Stetter home, the furnishings are contemporary. Heidi has combined warmth with open space. There is no clutter. Organized along an entire wall are volumes of family photos, which Heidi displays, revealing the different epochs of her family's life. One of the most memorable is of Stetter standing in the ocean near Vulcano Island in the Mediterranean Sea, 15 miles north of Sicily. The Romans named it Vulcano, the origin of the word volcano. Seems fitting he should be standing there holding a blue plastic tub filled with the first archaea samples collected from a hydrothermal vent; his small rubber raft in the background. Heidi, who has a Ph.D. in microbiology, and their five-year-old daughter, received samples of volcanic sediment from husband/father Stetter as he surfaced after many dives. This historic first collection of hyperthermophiles was a delightful family adventure.

Stetter is anxious for me to see the red Ferrari, where it would have to sit out the winter in the garage. Damn. It was beautiful. Most remarkably it was the product of success: forming an idea, developing a plan, and importantly, said execution of plan. I've seen Stetter beaming proudly on a number of occasions. The first time was when he showed slides on his laptop of his archaea. The second was an hour earlier when he introduced Heidi and showed pictures of his children. This was the third time. The fourth came moments later when Stetter led me to his greenhouse with his rare and exotic orchids.

"Cultivation, whether it is with these plants or with the microbes, is the same. You must know what they need for living, and not only for living, but for thriving. It must be the same for people too, do you believe? Otherwise we wilt and never see our blossoms," Stetter said as he inspected the under petals of a species of *Cattleya labiata*.

The next morning Stetter took me to the University of Regensburg. The snow had stopped. It was a clear, cold December day. There were rows of modern industrial-style concrete buildings with dark glass. These modern bunkers have become the world center of archaea research thanks to the guidance of Otto Kandler and Wolfram Zillig, who disciplined an extremely creative mind and introduced it to the stranger side of microbiology. Stetter was appointed full professor in charge of the department of microbiology in 1980. The labs were prepared to become the home of a type of microbial culturing system no one had seen before.

Stetter had been developing his methods for growing the sulfur-loving hyperthermophiles from Vulcano with Zillig at the Max Planck Institute. Now he would grow massive amounts of archaea in the laboratory at Regensburg, providing them to scientists around the world, just as Ralph Wolfe had done with methanogens. But before Stetter could begin mass producing microbes, workmen had to install his fermenting vats and hook up pipes to divert hydrogen sulfide gas made by the microbes to an independent ventilation system. As soon as that was done, Stetter poured his special ingredients into the vats and began brewing the stink bugs.

"I was walking to the laboratory in the next morning and I saw some students holding scarves and jackets to their noses," Stetter said, describing one of his early days on campus. "I became very nervous because I knew that something might be wrong with my laboratory. What had happened was the workmen had connected the exhaust for the hydrogen sulfide gas to the ventilation system for the entire building and everybody was breathing the gas from the stink bugs. It was not enough H_2S to harm them but they did not know this and they did not know what was causing the terrible smell."

After the ventilation episode settled down, Stetter got to the business of growing microbes. This also meant using high pressures and temperatures and lowering the pH to the lower extremes. The mixtures combined with the pressures caused severe pitting in the stainless steel wells of the fermenters in a relatively short time. Drawing upon his early background in engineering, Stetter came up with the idea for custom building enamel-titanium vats and also glass-lined steel to withstand the extreme conditions of his witch's garden. (See Plate 3.) But, the first test of the new system was not with sulfuric acids and seawater. "A local brewery filled a 300-liter vat with beer for a celebration, and then we began growing the stink bugs," Stetter says.

Stetter showed me the rows of blue fermenters cooking different species of archaea and led us to a small room with a peculiar-looking type of microscope with lasers attached. This was one of Stetter's greatest innovations in culturing archaea. He borrowed a technology developed by physicists for capturing particles, called "optical tweezers." Instead of particles Stetter isolated individual microorganisms with the optical tweezers for specific cultures.

According to a review published in the *PNAS* by Arthur Ashkin of Bell Laboratories, the ability to trap and manipulate small neutral particles by lasers is based on the forces of radiation pressure that arise from the momentum of light itself:[6]

> Nothing in the early history of light pressure forces using incoherent sources suggested useful terrestrial application. Only in astronomy, in which light intensities and distances are huge, did radiation pressure play a significant role in moving matter. With lasers, however, one can make these forces large enough to accelerate, decelerate, deflect, guide, and even stably trap small particles. This is a direct consequence of the high intensities and high intensity gradients achievable with continuous wave coherent light beams. Laser manipulation techniques apply to particles as diverse as atoms, large molecules, small dielectric spheres in the size range of tens of nanometers to tens of micrometers, and even to biological particles such as viruses, single living cells, and organelles within cells. Use of laser trapping and manipulation techniques gives a remarkable degree of control over the dynamics of small particles, which is having a major impact in many of the fields in which small particles play a role. In biological applications of optical trapping and manipulation, it is possible to remotely apply controlled forces on living cells, internal parts of cells, and large biological molecules without inflicting detectable optical damage. This has resulted in many unique applications.

This was fresh science when Ashkin published his review in 1997. Development of optical tweezers technology began in 1969. The first use in biology occurred in 1989 for examining the way bacteria move about with flagella, the bacterial "motors." The simplest explanation for what optical tweezers do is they work like a tractor beam. Opposing beams of lasers will cause particles to levitate, trapped by pressure forces, and then are able to be manipulated with the light beams. Stetter and his colleagues at Regensburg adapted the technology to create a laser microscope for individual microbes and published about it in 1995. More simple methods using solutions can be used for isolating microbes, but the optical tweezers are the most precise.

The laser microscope is able to separate a single cell from small colonies grown in nutrients. For the isolation of single cells, a cell separation unit was designed that consists of a 10-centimeter-long, rectangular glass capillary as an observation and separation chamber connected by a tube to the needle of a one-milliliter syringe. A cutting line separates the glass capillary into two compartments. After sterilizing the cell separation unit, about 90% of the glass capillary is filled with a sterile medium, Stetter's co-worker Robert Huber said. Afterward, the mixed culture is soaked into the remaining volume of the glass capillary. A single cell is selected under 1,000-fold magnification and is optically trapped in the laser beam. The cell can be separated within three to ten minutes at least six centimeters from the mixed culture into the sterile compartment by moving the microscopic stage. At the cutting line, the glass capillary is gently broken and the single cell is flushed into a sterile medium. This "selected cell cultivation technique" is safe, efficient, and fast and can be applied to many types of microorganisms, Huber explains.

Stetter left his program, albeit reluctantly, in the able hands of Huber, whose group continues to lead the scientific community in archaea research. After spending most of the day at the laboratory we drove back to his house for a home-cooked meal. Afterward, Stetter showed me a documentary from the 1990s about his work. We sat up most of the night talking about microorganisms and the origins of life.

"Now maybe you know a little bit more about the crazy archaea," Stetter said at the witching hour. I nodded gratefully and wondered how I will ever understand enough. On the flight home, the memory of an experience three years earlier began haunting me. I had seen archaea in action on the famous *Titanic* shipwreck, perhaps involved in a behavior reminiscent of a very ancient world, something that might even be occurring now on other worlds.

With my own eyes, I had seen the *thing*, a monstrous microbial wonder, but only now was I beginning to appreciate what it was—is.

5
RUST NEVER SLEEPS

As I peered down I realised I was looking toward a world of life almost as unknown as that of Mars.
William Beebe, 1930

Russian technicians have been working on *Mir I* and *Mir II* since 4 a.m., running diagnostics to ensure that the launch into inner space of the North Atlantic will be as successful as the previous 498 dives. From the open porthole of my stateroom, hints of dawn scattered into a dark, clouded horizon, hypnotic swells churned wisps of white foam on a rolling blue-black frigid sea. Foam fizzed at the crest of each wave. Rubbing sleep from my eyes, I sipped a little coffee and smoked a cigarette while the eastern sky slowly transformed into slate gray. Breaking the trance of first awakening, the engine of a 27-ton crane whirred to life on the working deck of the ship six flights below. Little knots twisted in my stomach. It was almost time.

Over fleece long johns I pulled on a pair of fireproof "nomex" coveralls, a size too small, issued from the crew chief. I filled a raggedy brown day pack with notebooks, pens, a digital camera, non-lithium batteries, a multi-purpose Nalgene bottle, and plenty of granola bars just in case. Lithium batteries are forbidden because of the risk of purple hydrogen gas fires. Small subs condense moisture so you don't want water dripping into a camera bag filled with lithium oxide.

In less than an hour, we'll be heading two and one-half miles to the bottom of the ocean, taking the same route made nearly a century ago by 1,523 men, women, and children who did not have the protection of a pressurized submersible. It is August 6, 2000. I've been aboard the *Akademik Mstislav Keldysh*, a 400-foot-long research vessel operated by the P. P. Shirshov Institute of Oceanology, for a week, but waiting a lifetime for this moment.

I'm the guest of treasure hunters who hired the *Keldysh* and its submersibles for a few million dollars for one month to make 28 dives to the *Titanic*. The stated purpose of the expedition was to retrieve artifacts—a salesman's kit with dozens of perfume vials, a workman's striped overalls, a child's math book, and hardware including two engine telegraphs, which were throttle-like stands used by the Captain on the bridge to signal the engine room to speed up or slow down. The artifacts were to be conserved for public exhibits, along with 5,000 other items collected from five previous expeditions by RMS Titanic, Inc., using a hired French submersible. But, the salvagers were actually hoping to find an alleged shipment of diamonds reportedly worth $300 million. They also intended to strip the deck of the bridge's telemotor that held the *Titanic*'s wheel, cut off one of the giant anchors, gut the hull to get at the ship's safe, and send robots into the first-class cabins to grab jewelry and valuables.

My stated purpose was to cover the expedition for the newspaper, presenting both sides of a flaming controversy: charges made by Bob Ballard and the National Oceanic and Atmospheric Administration (NOAA) of grave robbing and looting of a historic landmark, versus traditional rights to salvage shipwrecks under maritime law. Ballard, who led the American team that discovered the *Titanic* in 1985, along with NOAA, has been trying since 1986 to stop the salvaging of the wreck. This is a twisted and fated story, and the French have their hands deep in it. But with respect to all sides, fight among yourselves. The real reason I was there was to reach the bottom of the ocean and see the *Titanic*. Say no more. I would have come with terrorists had they invited me.

I had no inkling in August 2000 that three years later Eric Mathur and Karl Stetter would lead me on a global odyssey in search of archaea. But this expedition, more than any, seared my passion for the outer limits. As it happened, the *Titanic* became the focus of Roy Cullimore, a microbial ecologist who defies quick description, and Charles

Pellegrino, a Renaissance scientist and author in the vein of Carl Sagan. In 2000 it was common knowledge that bacteria were colonizing the shipwreck. But what was not appreciated then and still isn't now is that Cullimore and Pellegrino discovered that the *Titanic* is being devoured by the largest and strangest cooperative of microorganisms on Earth. We're talking one monster microbial industrial complex, as alien in its structure and productivity as it gets in the world of extremophiles. The corporeal rust comprises representatives of all three domains working closely together, constructing a slimy metal tissue and a vast circulatory system to nurture itself. It even developed an immune system of sorts.

Imagine mats of rust bigger than a dozen four-story brownstones that are alive, creeping slowly along the hull of the ship harvesting iron from the rivets and burrowing into microscopic layers of steel plating wherever they find a fracture, rip, or buckle. Consider 30-foot-long icicle-like growths, known as rusticles, dangling from the sides of the ship's bow. The mats and rusticles are a communicating super-organism, channeling iron-rich fluids, sulfur, and electrical power throughout the collectives of bacteria, archaea, and fungi. This is a highly organized entity at work in the pitch dark, near-freezing, low-oxygen waters of the ocean floor, beyond anything I could possibly make up. Sometimes a guy just gets lucky.

I wish I could tell you how I prepared extensively for this trip, researching everything known about iron-related bacteria and tracking down the scientists who published studies in obscure journals about the rusticles. But that's not how it happened. The assignment came quickly, and by chance I remembered a book sent as a review copy months earlier that was buried somewhere in a pile on my cluttered desk. I cracked the cover for the first time when my plane left Washington, DC. It was *Ghosts of the Titanic* by Charles Pellegrino. What I read gave me an entirely unexpected reason for wanting to see the *Titanic*:

> It was Roy who discovered, and began to map, an animal-like organization in those hollow spaces, who charted the paths of reservoirs, ducts, structural threads, and—could that be? Precursors to, or analogues of . . . arteries?
>
> "Look under the microscope and tell me what you see," Roy said.
>
> The . . . thing under the scope was more complex than a sponge, and a sponge was sufficiently complex to be claimed by the field of invertebrate

> zoology. The difference between the two was that a sponge cell, torn loose from its brethren, would die much in the manner of an ant separated from its colony, whereas the rusticle's cells were individual bacteria, and each bacterium could presumably survive on its own. What I began to see was the probable ease with which independently living, bacteria-like species might weave themselves into multicellular arrangements.

I wasn't sure whether to believe what I was reading. But the assignment to the *Titanic* had been nothing but surreal from the start. My connection to the expedition was Mike Manyak, then director of urology at the George Washington University School of Medicine in Washington, DC. Manyak is a surgeon by day, adventurer whenever possible, who recruited me into The Explorers Club in 1998. Manyak caught wind of the salvaging expedition one evening over martinis at the club's Upper East Side headquarters in Manhattan. More than several people in the club were stunned by what these "pirates" were planning to do to the *Titanic*. Salvaging shipwrecks for treasure hunting is ordinarily not an issue. But when it involved the *Titanic*, it became an entirely different matter.

Undeterred by the unsavory nature, Manyak contacted the treasure hunters. Being an experienced adventurer and board member of The Explorers Club, he suggested they acquire medical services for their expedition. One never knows what can happen in the middle of the North Atlantic and wouldn't it be comforting to have their own American doctors at the ready 24/7? The doctors would donate their medical services and supplies in exchange for, oh, say a sub ride to the *Titanic*. The salvaging company was under new management. The new leaders were mostly former concert and entertainment promoters; their staff, a Florida rock band. Having American doctors onboard in case a CEO had a heart attack or the band caught alcohol poisoning from the homemade "Sheila" distilled by the Russian crew sounded like a good idea. Plus, RMS Titanic, Inc., had an Explorers Club board member and attorney experienced at representing salvagers in court.

Manyak rounded up two docs, and being my pal, suggested to the treasure hunters that publicity in a big newspaper might do them some good. They thought so too. These particular pirates rarely got good press. One had a reputation for promoting dwarf-tossing contests at Florida bars. The other was being sued by stockholders amid rumors of an SEC investigation into his questionable takeover of the company. These guys offered me a ride to the *Titanic*, which I accepted with the understanding my articles would be fair to both sides of the controversy. I flew to St. John, Newfoundland, on July 28 to meet the pirates at a busy pier where they were loading supplies onto a decommissioned 140-foot British navy research vessel. The entrepreneurs bought the RS *Explorer* for transporting investors out and back for sub rides, and most importantly for getting booty off the *Keldysh* at the end of expedition.

The ship shoved off in the afternoon for a 26-hour excursion to the *Keldysh*.

That first evening I sat in the *Explorer*'s galley at a cramped table, surrounded by cases of wine, and gathered background information from the new leaders of the company, CEO Mike Harris and President Arnie Geller. They were rewriting company history and leaving out details about how they had taken over power in a stockholder coup from CEO George Tulloch the previous year. One would think they had conducted every expedition since the colorful history of the *Titanic* salvaging began. One important matter they glossed over was that Tulloch had been the heart and soul of RMS Titanic, Inc., since 1993, and was generally respected on both sides of the debate. From NOAA's perspective, he could have been worse. Tulloch agreed not to enter the ship or remove any structures from it. People who knew him said that even when no one was looking, he treated the shipwreck and the surrounding ocean floor where lost passengers lie buried in the sediments with respect.

Tulloch had taken over the company from another fellow, John Joslyn, who launched the first salvaging expedition in 1987 with members of the French team who co-discovered the *Titanic* with Ballard. This salvaging business started after Ballard's Woods Hole group tested an experimental remotely operated vehicle (ROV) called *Argo*, equipped with a camera and developed for the U.S. Navy to secretly locate enemy submarines. The Institut Français de Recherche pour l'Exploitation Durable de la Mer (IFREMER), led by Jean Louis Michel, was the French partner in a classified joint exercise. The French were testing their new sonar device, systematically scanning the ocean floor, while the Americans deployed the ROV and looked around.

Neither group was there officially to search for the *Titanic*, but since they would be in the North Atlantic, why not move a little closer to where the *Titanic* might be resting? Three unsuccessful attempts in the early 1980s by Texas oil millionaire Jack Grimm using state-of-the-art sonar told the American/French team where the *Titanic* would *not* be found. During the month-long exercise the French team further eliminated wide swaths of ocean to narrow the search. A month into the formal exercise, the French returned to port. Ballard's group stayed on and kept looking with *Argo*. They located the *Titanic* three weeks later.

Not long after Ballard's discovery, rumors circulated that IFREMER might disclose the location of the *Titanic* to treasure hunters. Around this time Ballard and NOAA got Congress to pass the R.M.S. Titanic Memorial Act of 1986 to declare the site an international memorial to the dead and to state no one be allowed to salvage the ship. But under maritime law, whoever claims a wreck as salvor and brings back proof of the find, and demonstrates the ability to return to the wreck to get more booty, owns the salvage rights. Ballard's group did not claim salvage rights, figuring he could keep the location secret.

To Ballard and NOAA's horror, Joslyn's group and the IFREMER's *Nautile* made the first dive in 1987 to retrieve artifacts. Joslyn's group filmed part of the event live with Telly Savalas hosting from Paris. Tulloch, a former car dealer from Connecticut, took control of the salvage rights May 4, 1993, from Joslyn's Titanic Ventures Limited Partnership. The following month, RMS Titanic, Inc., and IFREMER conducted a second joint expedition to the wreck site. They made 15 dives and recovered 800 artifacts.

NOAA and Ballard continued challenging salvage rights, but in 1994 a U.S. Federal Court granted formal "salvor-in-possession" to Tulloch's company. That summer Tulloch's group and IFREMER returned to the wreck site and completed 18 dives, recovering 700 artifacts. The court order was reconfirmed in 1996. That year Tulloch arranged for 1,600 people onboard the cruise ships MV *Royal Majesty* and SS *Island Breeze* to join the expedition. Tulloch hit all the right buttons, bringing along *Titanic* survivors Michele Navratil, Edith Haisman, and Eleanor Johnson Shuman. Tulloch easily convinced Discovery Channel to pay $3 million to film a documentary on the *Titanic* and on the raising of a 17-ton section of the hull. IFREMER's ship *Nadir* and the submersible *Nautile* were again hired by Tulloch along with the *Ocean Voyager*, an oceanographic research ship. Cullimore and Pellegrino were invited to the 1996 event when they began the first dissections of rusticles.

"I went down to the ship in 1996 and did a survey of the promenade deck and yes there is definitely life down there. I had an idea about how to detect the presence of microbes very quickly so I left some color slide film on the bow. The bacteria get into the slide film and mine the gelatin, leaving beautiful colors behind that look exactly like stained glass windows when it is processed," Cullimore said.

A 1996 Associated Press account by science writer Matt Crenson characterizes a wild atmosphere during the 1996 expedition, and inadvertently describes the beginning of rusticle dissection begun by Cullimore and Pellegrino:

> Two cruise ships, charging thousands of dollars per cabin, will come alongside the expedition's research vessels next week as they attempt to lift a 400-square-foot section of the Titanic's hull to the surface. Three survivors of the April 14, 1912, shipwreck will be present, as will actor Burt Reynolds and former astronaut Edwin "Buzz" Aldrin. . . . The Discovery Channel paid R.M.S. Titanic $3 million for exclusive rights to film and photograph the recovery of the hull section [which failed], as well as the month long expedition that led up to it. In the last few weeks, the French submersible Nautile has brought serving dishes, rusty metal scraps and bottles of Bass Ale from the wreck back to daylight. About 12,000 bottles of Bass were thought to have gone down with the Titanic, and the 10 lucky winners of the Bass Ale Voyage to the Titanic sweepstakes were there to see a 1912 vintage of their favorite brew brought to the surface.

The "rusty metal scraps" were the rusticles. Microbiologists Henrietta Mann and William Wells at St. Mary's University in Halifax would later identify at least 20 different bacterial species, 2 fungal species, and at least 1 species of archaea.

In 1998, Tulloch returned with *Nautile* and the French ship-recovery vessel *Abeille Supporter*, with a successful recovery of the huge hull section. The Discovery Channel was back again. NBC's *Dateline* aired live segments, and Cullimore returned to follow up on rusticle research, which was revealing startling new information.

"We returned in 1998. We discovered that when metals twist and become brittle they become home to so many microbes. The engineers just ignored us, and so did just about everyone else. When they finally brought a big section of the hull up in 1998, the engineers were breaking off the rusticles and tossing them overboard as if they were junk. But we, the microbiologists, did more to contribute to the history of the decline of the *Titanic* than anyone," Cullimore said.

Had Tulloch remained as CEO of RMS Titanic, Inc., Cullimore

would have returned again in 2000. While the microbial research in 1996 and 1998 was overshadowed by engineers' theories of how the *Titanic* sank, Cullimore said Tulloch strongly supported continuing rusticle research. Pellegrino wrote that Tulloch had been interested in the rusticles at least since 1991, and by 1995 secured the scientific expertise of Cullimore. Tulloch introduced Pellegrino and Cullimore in 1996. Needless to say they clicked. It was clear from what I had read so far in Pellegrino's book that he and Cullimore respected Tulloch as a decent steward of the *Titanic*'s salvage rights, despite the publicity circus.

The new guys were different. They weren't interested in science. They immediately regretted having me along to report on their expedition. Over dinner on the *Explorer*, they talked about the diamonds, the probable location of the safe, ways to penetrate the ship, and how to cut off the *Titanic*'s portside anchor. They hired a second vessel to operate two ROVs equipped with robot arms to load as much stuff as possible from inside the *Titanic* into baskets big enough to lift a car. It was strange hearing these guys talk as if they owned the *Titanic*, which in a sense they did, and could do whatever they wanted, which in a sense they could. Although, in the end a federal court order issued two days into the expedition banned RMS Titanic, Inc., from entering the wreck and cutting anything off the hull and deck. Secondly, the heavy seas proved too much for the flat-bottom ship hired as a platform for the ROVs. They were out of commission after a few days. Thankfully, this expedition will not go down in history as the one that finished off the *Titanic*. The new leadership was interested in talking about the rusticles, but only insofar as the rust advanced their agenda to strip the *Titanic*. They did not care about the scientific insights into microbiology. Their talking points for the media, which was me, were simple. The *Titanic* is decaying rapidly. Microbes are accelerating the rate of decay. Therefore, they should be allowed to recover as much stuff as possible while the ship remains relatively intact.

CEO Harris told me, "If you take the long view, don't you think future generations will look back and ask why didn't you save those things when you had the chance?"

Harris and Geller intended to retrieve artifacts from the mailroom, first-class cabins, and officers' quarters, and anything of historic value along with money, gold, and jewelry. They planned to remove the bronze-covered tops of capstans on the bow used for tying up the big mooring lines in port. No doubt the public would clamor to see all of these things in an exhibit. But based on what I was reading in *Ghosts of the Titanic* this didn't sound at all good for the ship. Any bump by a submersible, any scratch in the surface, especially cutting straight through the hull and pulling down cabin walls, would be like tilling a garden for the aggressive microbial consortium.

Cullimore had been calculating the rate of decay based on the growth rates of new rusticles he was growing on steel plates placed on the sea bed in 1996. It was clear from videos taken by Tulloch's group and the French since 1987 that the rusticles on the ship were spreading. Creating new damage would open portals for the microbial monster to insert new tendrils. Frankly, I was feeling horrified. Cullimore and Pellegrino were calling the *Titanic* "Rusticle Park" for the aggressive behavior of the microbial consortium. We should not be tilling it, feeding it.[1,2]

In my cabin that night I clicked on a dim reading lamp above my bunk, pulled on a thick blanket to keep away the Atlantic chill, and as the *Explorer* pitched and rolled in the swells at about 10 knots, I read more: "Huge sections of the promenade deck were being devoured. When Bob Ballard made his first reconnaissance of the bow in 1986, only the promenade windows were missing. When Roy Cullimore photographed the same section in 1996, rows of vertical supports had dissolved almost out of existence, and by 1998 a forty-foot length of the boat deck had collapsed into the promenade. Stronger. The rusticles appeared to be getting stronger, and hungrier," Pellegrino wrote.

Roy Cullimore, Ph.D., is a slender, sandy-haired, bearded professor of microbiology at the University of Regina in Saskatchewan, Canada, now retired, and president and CEO of Droycon Bioconcepts Inc. Cullimore spent the past 45 years studying the personal habits of iron-related bacteria and the groupthink of microbial communities. Cullimore has thoughts that don't occur to many people. He perceives the world quite differently from most of us. For one, many years ago Cullimore determined that clouds are giant biological organisms. When he looks at the sky he sees a biocosm. When he sees lightning, he's watching electrically charged bacteria bump and grind. With Cullimore's microbial clouds, we're not talking about pathogens that get swept up in windstorms and transported from Africa to the Caribbean. Cullimore means that colonies of electrically charged bacteria are the essence of clouds; and at least one cloud variety has specifically evolved to form ice crystals that induce cloud formation. Put enough of these microbes in a bucket of water and you can make ice at 55 degrees F. You can try this at home with *Pseudomonas syringae.* They're all over plants. This is old news. Freeze-dried cells of these microbes are commonly used to create artificial ice islands for offshore oil drilling in the Arctic, and they make artificial snow for ski resorts. But they're so cool the Department of Energy is sequencing the genome of *Pseudomonas syringae* to get down to business with these organisms. Industry has ideas about making artificial mountains of microbial ice in the winter for cooling office buildings and warehouses in the summer. Imagine that in your air conditioning.

But something I had not heard before from Cullimore was that Mars might have rusticles. Cullimore said he is certain he saw a rusticle in one of the first photos transmitted by *Spirit* from the Martian surface. Perhaps there are consortium-like, iron-loving entities on Mars or consortiums of other microbes on other worlds engaged in all varieties of strange biochemical industries. This is not as crazy as it sounded 10 years ago.

Cullimore isn't one to allow dogma to chase his thoughts away. It didn't strike him odd in the least the first time he saw a picture of a rusticle and wondered if the rust was alive. Ballard coined the term rusticle because the peculiar structures look like rust icicles. To Cullimore, the rusticles were awfully similar to the clumpy formations that colonies of iron-loving bacteria create in deep freshwater wells. Cullimore suspected microbial mischief afoot on the *Titanic.* He has been trying since the 1960s to outsmart the cagey bacteria that eventually plug most wells. He also specializes in keeping iron-loving bacteria from clogging water pipes, corroding oil pipelines, and pitting the cooling systems of nuclear reactors.

One thing that puzzled me was, how do microbes always seem to show up in extreme environments ready to carpe diem? When new hydrothermal vents open on the cold, dark, ocean floor, where do the hyperthermophiles come from? Are they floating in some dormant state waiting for a gush of boiling sulfuric acid to wake them up or are they already down in plumbing systems below the seafloor in the crust and do they just happen to be ejected through new vents? What about the bacteria, archaea, and fungi colonizing the *Titanic*? How did they find the *Titanic* when it took us more than 70 years using secret navy technologies? Were the iron-loving microbes already at work on the new steel before the ship set sail? Were they already present in the ocean scratching out a meager existence from molecules in the bottom sediments? Pellegrino pondered whether microbes from the toilets on the *Titanic* and the throats of passengers survived and mingled. When I posed the question to Cullimore he quoted Dutch microbiologist Baas Becking:

"Everything is everywhere. The environment selects."

Becking was summarizing the two fundamental laws that govern microbial behavior, proposed in the late 19th century by the "true father of modern microbiology," Martinus W. Beijerinck.

What Cullimore, Pellegrino, and their colleague Lori Johnston are still trying to figure out is how the microbes developed into the monstrous rust entity. How did members of three domains become so communal? Are organization, cooperation, and communication programmed into all microbial genes? If so, it seems nature selected microbes, predetermined them, if you will, to organize into ever more complex structures under certain circumstances, perhaps even into multi-cellular organisms given time and circumstances. If true, what are those circumstances and where might we look for other examples? Cullimore insists the rusticles are using a common microbial language, which could be chemical or even electrical. Microbes engage in a phenomenon called "quorum sensing" through which whole communities "sense" each other's presence and activities and respond appropriately with chemical signals to divide, share resources, circle the wagons, or recruit allies depending on the circumstances. Microbes even sacrifice themselves in hard times for the greater good of the collective in a form of microbial Marxism. Cullimore believes quorum sensing is aiding and abetting the organization, cooperation, and growth of rusticles. He and Pellegrino see a form of social "intelligence" at work inside the *thing* that is harvesting the *Titanic*.[3]

Quorum sensing is an incredibly important and fascinating area of microbiology. Infectious bacteria engage in this behavior inside our bodies. They invade but do not attack until they "sense" that they have reached sufficient numbers to overwhelm our immune systems. To find a gargantuan microbial entity like that of the rusticles offers a rare opportunity to study microbial life at a truly strange and opportunistic moment in time. It must hold many secrets to the past and future of life on Earth.

I was ready to see the rusticles up close. We boarded the *Keldysh* in the late afternoon after a cold, wet ride on a rubber dingy sent in high swells from the *Explorer*. The *Keldysh* is a grand research vessel built in Finland for the former Soviet Union and launched in 1981. The *Mir* submersibles, also built by the Finns, were christened in 1986. Their interiors are the size of a *Soyuz* space capsule, slightly larger than our old *Apollo* modules. The diameter of the *Mir* cabin is seven feet. Each sub carries three people to depths up to 20,000 feet.

You may be familiar with the *Keldysh* and *Mir* submersibles already. James Cameron employed them to film the movie *Titanic* and the 3-D IMAX documentaries *Ghosts of the Abyss* and *Aliens of the Deep*, which was about hydrothermal vents. The white stern railing of the *Keldysh* is where *Titanic* movie character Rose tosses the big blue diamond into the sea at the end. After the Soviet Union collapsed in 1989 and the *Keldysh* and *Mir*s lost state funding, films kept the *Mir*s diving and more recently the *Keldysh* and *Mir*s went to work servicing oil rigs in the Baltic Sea, where they are today.

After settling into a spacious cabin on the officers' deck of the *Keldysh*, I scramble down a metal stairway to the briefing room on the *Keldysh*'s main deck. (My luxurious digs were secured by Ralph White—one of the original discoverers of the *Titanic* and the man who introduced Cameron to the *Keldysh*. I met him on a previous expedition with submersibles in California's Channel Islands.)

Harris, the CEO, arranged for me to dive with him and Anatoly Sagalevitch, director of the MIR submersible research program for the Russian Academy of Sciences, arguably the most experienced and able submersible pilot on Earth. Harris wanted me to observe how rapidly the *Titanic* is disintegrating to boost his chances of overturning the court order against salvage operations. This guaranteed our dive together would focus on surveying the rusticles rather than salvaging!

The 25-foot-long *Mir*s sit one in front of the other on elevated, steel-grated platforms in hangers near the center and starboard side of the *Keldysh*'s rear deck. The crane is positioned between them. The submersibles are launched one at a time and always travel as a pair. If one sub is disabled the other can lock onto it with its robot arms and the twins can ascend together to the surface. If one gets stuck, beneath an overhang, for example, which has happened, the other can use its robot arms to remove the obstruction. Anatoly finished the briefing and signaled for Harris and me to follow him. We climbed a ladder to

Mir I and removed our shoes to prevent tracking flammable grease into the cockpit. (See Plate 4.)

We slipped through the hatch. Anatoly had exactly enough headroom to stand erect in the center of the sub below the hatch with an inch of space above his head. The pilot sits upright at a center console with the instruments, a sonar screen, and joysticks for operating the thrusters and manipulating the dual robot arms at his fingertips. His view port is seven inches in diameter. The two passengers can sit upright on padded benches on the port and starboard sides of the cockpit or lie comfortably prone to peer through the four-inch-diameter view ports positioned forward and slightly to the sides like the eyes of a frog. Harris took the starboard position. I settled in at port. Next to me were gauges for monitoring carbon dioxide levels, which we recorded every half hour. The sub is a tightly contained environment pressurized with the pure oxygen we breathe. Scrubbers—alkaline filters consisting of soda lime—absorb carbon dioxide from the re-circulating cabin air. The scrubbers last for about three days before they clog with carbon dioxide. If the submersible became disabled and rescue was impossible during this critical window, drowsiness leads to sleep and then death.

After the hatch was closed, the cabin was pressurized with a hiss by O_2 to equal the normal ambient pressure of one atmosphere outside at the surface. Technicians simultaneously locked a thick cable on to the top of the sub. The crane has two crab-like grips secured over the upper sides of the sub, which the whirring crane hoists over the side of the ship and lowers into the water by slowly letting out the cable. A tender—a hefty 30-foot motor boat launched first—positioned itself about 50 feet from the subs once they were in the water, then it launched a motorized dingy with a three-man crew wearing heavy wetsuits. One of the men must jump from the dingy into the 40 degrees F water, climb on top of the sub, and physically uncouple the cable. Inside, we felt ourselves being swung out by the crane and lowered into the water where we bobbed like a cork in the swells. After a few minutes we heard the diver hopping onto the sub; his wetsuit squeegeed against the hull as he grasped the cable. Then clank. We're free. We pitched about for another 10 minutes while *Mir II* was lowered and released. We drifted safely away from the ship with the current.

Anatoly waited for topside to tell us to dive. The radio crackled to life, something from a 1950s science fiction movie with reverb and static. The conversation between Anatoly and topside was made more exotic spoken in Russian. Anatoly nodded. We could dive. He recorded our longitude and latitude at the surface then released air ballast from tanks below the submersible. We descended in a stream of bubbles. At about 80 feet per minute, the trip took about two and one-half hours. Because of thick clouds overhead, the first 300 feet of the surface was a shade of deep blue that does not exist in a box of crayons. If the sun had been out it would have a greener tint. This was the epipelagic zone, the sunlit upper layer from the surface to 300 feet deep. Much of the critical action for life as we know it happens there, but it is only the first link in a chain that reaches all the way to the abyss.

The epipelagic zone is biosphere Earth's primary life support system, powered by sunlight and manned by industrious cyanobacteria and microscopic algae, which are the phytoplankton. They contain chlorophyll and use photosynthesis to convert carbon dioxide diffused from the atmosphere into carbohydrates. The primary nutrients for phytoplankton are nitrogen, phosphorus, and iron,

Phytoplankton are the primary CO_2 scrubbers and O_2 generator for the "pale blue dot" called Earth. The development of photosynthesis by cyanobacteria is thought to be the event that converted Earth from archaea's carbon dioxide haven into the oxygenic atmosphere it is today. Phytoplankton are more important than plants and trees in scrubbing CO_2 and supplying O_2. Oceans cover two-thirds of the planet's surface. Do the math. If the phytoplankton were wiped out or severely depleted, the CO_2 they normally absorb would remain in the atmosphere, accumulating and accelerating warming. All the trees and plants in the world could not absorb enough carbon dioxide and produce enough oxygen to save us. Ultimately, we would become drowsy, then sleepy, and die like passengers in a disabled submersible. Microorganisms that abhor oxygen and love CO_2, archaea and bacterial anaerobes, would dominate again as they did on Baby Earth. Eventually the atmosphere would consist mostly of CO_2 and a little nitrogen. Atmospheric pressure would increase 60-fold. Air temperature would rise to nearly 600 degrees F. We would become like Venus.

We may have started down that path. NASA satellite studies led by Watson Gregg documented a drop in phytoplankton production from 1980 to 2002. His team measured net primary productivity, which they define as the rate at which plant cells take up CO_2 during photosynthesis. Globally, the rate of CO_2 uptake declined 6%. Nearly 70% of the global decline of phytoplankton per decade occurred in the North Pacific and North Atlantic. Phytoplankton loss is occurring due to a combination of events. As phytoplankton productivity falls, ocean temperatures are warming. Warmer surface temperatures prevent nutrients at the surface from descending and nutrients accumulated on the bottom from cycling upward. Whatever the reasons—natural cycles, man-made pollutants, or both—the sunlit layer of the ocean is warming, nutrient levels are falling, and phytoplankton activity is down globally.

Until around 2000, scientists assumed phytoplankton consisted of algae and some bacteria. But microbial ecology's sampling techniques using raw DNA revealed that bacteria and archaea are thick as thieves in the epipelagic zone. Not much is known yet about their role but scientists worldwide are working on that. What is becoming clear is microbes in the sunlit layer are responsible for trapping about half of the carbon dioxide formerly attributed to algae. The dominant microbes in the sunlit zone are called picocyanobacteria because they are so tiny. These little guys are of tremendous ecological importance. The Department of Energy's Joint Genome Institute (JGI), The Institute for Genomic Research (TIGR), the Venter Institute, and the European Initiative involving Genoscope, the EU Margenes project, NoE Marine Genomics Europe, and the Pasteur Institute began sequencing the genomes of marine picocyanobacteria in 2003. The international project is sequencing 11 *Synechococcus* strains and about a dozen *Prochlorococcus* strains. They are the two most common types of marine bacteria, accounting for most of the microbial photosynthesis in the oceans. They scrub about 10 billion tons of carbon from the air each year. Pretty amazing for something no one could culture or knew existed a decade ago.

Prochlorococcus marinus may be the most abundant photosynthetic cell on Earth. A single milliliter of seawater holds more than a million individuals. Think about that the next time you gulp seawater while swimming. *Prochlorococcus* inhabits tropical and temperate oceans. Science writer John Whitfield of *Nature* wrote about a strain found in the Mediterranean Sea and sequenced in 2003. It has only 1,700 genes, one of the smallest genomes found so far. Another species from the Sargasso Sea has a similar-sized genome. The common green alga *Chlamydomonas* contains 50 times more DNA and is 15 times larger in cell size than *Prochlorococcus.* In the low nutrient regions of the ocean where the algae industry is normally small, such as in the Sargasso Sea, the *Prochlorococcus* genus contributes up to 40% of the biomass of phytoplankton.

Two strains of *Synechococcus,* WH7803 and RCC307, were sequenced in the same project. One of these has a genome of 2,500 genes. According to Whitfield's article, *Synechococcus* is less common, found in concentrations of about 10,000 cells in a milliliter of seawater. But it is more widely distributed in the oceans than *Prochlorococcus. Synechococcus* is able to work with a wider range of elements than *Prochlorococcus. Synechococcus* possesses enzymes adapted to metabolizing nickel and cobalt, perhaps to conserve iron. Its versatility is further aided by the ability to tap multiple sources of nitrogen. "It seems to be more of a generalist than *Prochlorococcus*—it's good in a range of environments," said Brian Palenik of the Scripps Institution of Oceanography, who led the *Synechococcus* sequencing team.

More recently discovered is *Pelagibacter ubique,* one of the most ubiquitous bacterial species of the oceans (hence the name) belonging to the very big bacterial group known as SAR11. Teams at the University of Oregon, University of Hawaii, JGI, and Diversa found *Pelagibacter* has the most compact, streamlined genome of any *free-living* microorganism discovered so far. With only 1.3 million base pairs, it weighs in as the smallest known microorganism yet. Stetter's *Nanoarchaeum* is actually twice as small, but as a symbiont lacking its own metabolism it isn't a contender for the free-living division. *Pelagibacter* has a complete set of biosynthetic pathways and can reproduce efficiently by consuming dissolved organic matter in the water. Its neatest evolutionary trick is having the ability to make energy from next to nothing. It is found all over the oceans, but like its buddy *Prochlorococcus* it dominates the most nutrient-deprived regions, including the Sargasso Sea. If nutrient levels continue falling in the sunlit layer, it's possible *Prochlorococcus* and *Pelagibacter* could keep the CO_2 and O_2 industries running for a while. How long and how much CO_2 they can scrub is anybody's guess. But scientists theorize these microorganisms might have evolved to become pinch hitters when A-team algae are too fatigued to keep taking CO_2 pitches and batting O_2.

Bacterial ecology of the oceans is becoming a dynamic field of study, powered by the machines of the world's largest gene sequencing centers and absolutely dependent on a handful of scientists' appreciation for the importance of microbes in the sea.

It appears now that microbial diversity of the ocean is up to 100 times greater than expected. The vast majority of species are unknown. This conclusion was drawn from research published in July 2006 in the *Proceedings of the National Academy of Sciences* by a team led by Mitchell L. Sogin, director of the Marine Biological Laboratory's Josephine Bay Paul Center for Comparative and Molecular Biology and Evolution, located in Woods Hole, Massachusetts.[4]

"These observations blow away all previous estimates of bacterial diversity in the ocean," Sogin said. "Just as scientists have discovered through ever more powerful telescopes that stars number in the billions, we are learning through DNA technologies that the number of marine organisms invisible to the eye exceeds all expectations and their diversity is much greater than we could have imagined."

We already know fewer than 6,000 microbial species have been identified based on cultures. "This study shows we have barely scratched the surface. Over the last ten to twenty years, molecular studies have shown there to be more than 500,000 kinds of microorganisms. In our new study, we discovered more than 20,000 in a single liter of seawater, having expected just 1,000 to 3,000. The number of different kinds of bacteria in the oceans could eclipse five to 10 million," said Sogin.

The study was part of the International Census of Marine Microbes, a project under the umbrella of the Census of Marine Life, a 10-year global initiative started in 2000. It now involves more than 1,700 researchers in 70 countries. The goals are to survey and understand diversity, distribution, and abundance of life in the oceans in the context of the past, present, and future.

Sogin's team used revolutionary new technology for sequencing raw DNA called "454 tag Sequencing." The most unusual sequences of microbial DNA belonged to organisms estimated to have very small populations. The 454 tag Sequencing technology is the next evolution in gene sequencing, accomplishing in three days what earlier had taken a month.

"The detection of these previously overlooked microbes opens a world of new questions about their role in ecological processes and their evolutionary history," said Sogin. "Exploration of this newly discovered 'rare biosphere' could become a major field of marine biology. We know we're going through global change, and microorganisms are vital to our survival. What we do not know is the role of low-abundance microbial populations that have persisted over large evolutionary time scales. We need to understand their diversity and how they work, to anticipate which species can be expected to adapt and how quickly, and to grasp how such other forces as competition and resource availability drive evolutionary change."

Large populations of a small number of species usually dominate environmental samples. But thousands of low-abundance species represent the overwhelming majority of unidentified sequences in Sogin's study.

"The most intriguing immediate questions are: How and why do these many different populations of low-abundance organisms exist at all; why are they still here? It's possible these rare organisms are present in high numbers at some locations and their low abundance at other sites is a consequence of diffusion and dispersal. However we believe they must have some importance to the marine ecosystem and can theorize a couple of possible roles."

One possibility, he said, is they are "keystone life forms" that play an important role within a microbial community, producing perhaps an essential compound. Another is low-abundance organisms are nature's backup players in the game of natural selection. Perhaps there is safety in small numbers. Operating below the radar of predators they don't have to compete with dominant species. They might exist throughout the oceans, able to pitch in if environmental conditions suddenly change and the presumed dominant bacterial society goes into a tailspin.

"In other words, the low-abundance organisms might not be very active but represent a reserve of genetic and genomic innovation," said Sogin.

The National Science Foundation has been funding research since 1988 at the Station ALOHA, Hawaii Ocean Time-series (HOT) research site. Researchers there have cataloged information about the biological, chemical, and physical parameters of the surrounding ocean, making it one of the most comprehensively characterized sites in the world. Research led by Ed DeLong, now at the Massachusetts Institute of Technology, sequenced communities of microorganisms at depths from 40 feet to 13,000 feet, then compared genomes of the different communities. This revealed the specialized roles of microbes from the surface to the bottom.

"By reading the information stored in the genomes of entire microbial communities, we can begin to measure the pulse of this marine ecosystem," DeLong said. "These new DNA sequences from microbial communities will help us paint the picture of how that world works and provide important details on the players involved and their biological properties and activities."

The University of Hawaii team sequenced 64 million base pairs of DNA from microbes and viruses collected at each depth. HOT discovered thousands of new genes from unknown microbes and found evidence of frequent gene exchange between organisms. Variations in genetic composition exist at different depths including differences among genes involved with carbon and energy metabolism. Gene swapping is an ancient behavior with highly adaptive benefits for species undergoing environmental stress.

"Although they're small, these tiny microbial species are the engines of the biosphere and in large part drive the cycles of matter and energy in the sea," DeLong said. HOT research published in *Science* showed microbes in the sunlit zone had more genes devoted to taking in iron, a key nutrient for phytoplankton. Deep-dwelling microbes contained DNA associated with "jumping genes," pieces of DNA that can move from one part of the genome within a microorganism to another. Jumping genes in deep ocean microbes might enhance their adaptability to changing environments, allowing them to tap different reserves of nutrients and alter their metabolisms when necessary.

In the food chain of the epipelagic zone, fish larvae and baleen whales thrive on phytoplankton while microscopic animals—the zooplankton—rise from the deeper ocean at night to graze on carbon-fatted algae and bacteria. Below the sunlit layer, *Mir I* descended quickly to the mesopelagic layer, also called the twilight zone, which extends from 300 feet down to 3,300 feet. This dark region is home to trillions of phosphorescent copepods, which constitute the zooplankton. The International Census of Marine Life database has logged 7,000 species of zooplankton so far. Higher up in the twilight zone's food chain are jellyfish ranging from the size of thumbtacks to the size of a barn. Here are shrimp, squids, octopi, and all manner of exotic fishes. Lantern fish resemble ferocious sea monsters except they fit in the palm of your hand. I held one on an unusually balmy midnight sampling cruise aboard *Keldysh* with Russian biologists.

The mesopelagic zones of the Atlantic, Pacific, and Indian Oceans and deep fiords are normally abundant with strange fish. These fish detect photons invisible to our eyes and many have light organs to lure prey, scare predators, and woo mates. Most life in the twilight zone generates its own light.

From the beginning of the descent in *Mir I*, I was prone, eyes glued to the thick view port. At about 500 feet, my eyes no longer detected photons except for the reddish ambient light of the cockpit. The temperature was quickly falling inside the sub, but I pulled my jacket off and covered my head to watch luminescent jellies and thousands of creatures fluoresce in response to our invasion. Peering downward, the flashes reminded me of stars blurring when the Starship Enterprise launches into hyperspace. Some species make their own luminescent fluids, but the majority employ bioluminescent bacteria in a symbiotic partnership. These bacteria are used in laboratories to tag DNA to make proteins and expressed genes glow.

This is part of what marine biologists call the "daily commute." After sunset, bioluminescent zooplankton and mid-ocean fish migrate up from the twilight zone to feast on phytoplankton, picoplankton, and fish larvae in the epipelagic layer. Waving your arm creates a luminescent trail. Just before dawn, zooplankton migrate back to their netherworld, where they become the main course for larger species of the mesopelagic zone. Zooplankton comprise a significant portion of "the deep scattering layer," discovered in 1948 when naval engineers were testing sonar and kept getting interference. The deep scattering was later found in every ocean except the Antarctic. It is a veneer of mesopelagic microbes and fish that covers most of the planet.

Something so global must have a starring role in the cycle of life. Zooplankton ecology has a key part in carbon cycling in the ocean. Zooplankton and phytoplankton production are intimately linked. Recall NASA's observation of the 6% global decrease in net primary productivity, the rate phytoplankton cells take up CO_2 during photosynthesis? The net primary productivity is an average. Detailed data show a 7% decline in the North Atlantic basin, a 9% decline in the North Pacific basin, and a 10% decline in the Antarctic basin. Cullimore and Pellegrino suggested nine years ago that overgrazing of phytoplankton by zooplankton was occurring. The reason, they said, was zooplankton populations were exploding because of overfishing by humans. Fish are the primary predators of zooplankton. The decline of phytoplankton productivity is happening in regions of heavy commercial fishing. Less fish = more zooplankton = less phytoplankton = more CO_2 = less O_2. This is actually tied to the accelerating biodegradation of the *Titanic.* The ship is a canary in the cycle of ocean life.

The bioluminescent universe of the twilight zone began to fade and the twinkling all but ended as we descended to 5,000 feet. Photons still penetrate at this depth but not many. There is, however, another light show that takes place. If I had ultraviolet vision I would have seen a shimmering iridescence, a silvery aurora. Fish and transparent jellies would have appeared silhouetted against this background of UV light. Some predator species down there possess UV vision.

The deeper we traveled, the more microbial life dominates as biomass. At the time of my dive, scientists assumed archaea were restricted to extreme environments. But since then it's become clear that archaea play an important role in the open ocean from the surface to the abyss along with bacteria. Archaea were found in the sunlit layer with sensors used to chemically convert sunlight into energy, similar to halophiles. These sensors capture solar radiation and turn on genes to manufacture a substance called proteorhodopsin. According to research published in the *Public Library of Science*, bacteria with proteorhodopsin may be using the solar energy to metabolize sulfur, a common energy source for deep sea life. Genetic studies also suggest marine bacteria can manufacture retinal, a molecule typically associated with vision. "Many other organisms use proteins resembling proteorhodopsin for different functions. Humans, for instance, use rhodopsin to sense light in the eyeball. The presence of rhodopsin-like proteins in a wide range of life may eventually provide hints to the protein's evolutionary age," the study said. The salt-loving archaea found in brines, commercial ponds, the Dead Sea and Great Salt Lake, develop purple membranes with sensors that stimulate proteorhodopsin for chemosynthesis of sunlight.

From the twilight zone to the ocean floor, estimates of archaea's biomass range from 40 to 80% of the total depending on where the sampling is conducted. One study found archaea constitute 20% of all picoplankton of the world's oceans. *Prochlorococcus*, *Synechococcus*, and

Pelagibacter are all bacterial picoplankton. As the names suggest, they are tiny. Picoplankton include bacteria, archaea, eukaryotes, and ocean viruses. A drop of seawater will hold a million or more of these microorganisms. Less than a decade ago, no one knew they existed yet they are the most abundant ocean life-form, no doubt playing a substantial role in the uptake of carbon dioxide and production of oxygen for Earth's biosphere.

In August 2000, David Karl and Markus Karner of the University of Hawaii, and Edward DeLong, then at the Monterey Bay Aquarium Research Institute, were wrapping up a study of deep ocean microbes in the Pacific. They were the first to find archaea in the open oceans. It consisted of species that did not fit the normal extremophile definition. In a paper they published in *Nature*, the scientists wrote: "As a dominant component of the ocean, archaea are thus far from confined to extreme niche habitats. Rather, the distribution of these archaea suggests that a common adaptive strategy has allowed them to radiate throughout nearly the entire water column." Archaea found in surprising abundance in soil contain a substance known as crenarchaeol, which allows them to adapt to milder temperatures. Something similar may be afoot in the open ocean. How remarkable that a form of life ignored and ridiculed by scientists less than two decades ago turned out to be one of the most predominant and adaptable forms of life on Earth. (See Plate 4.)

The idea of archaea as an extreme form of life needs revising since the recent discovery of abundant temperate species in the oceans, soils, and freshwater ponds and lakes. Archaea have a distinct survival advantage. They have had at least 3.5 billion years to work on it. They have adjusted comfortably to our narrow habitable zone, but we don't stand a chance in the hell of their outer limits.

More recent research led by Karl and Karner used the 16s rRNA method to conduct gene surveys from the mesopelagic to the abyssal zone. They sampled seawater monthly for a year and found two specific archaeal groups, the Euryarchaeota and the Crenarchaeota, thriving in the deep blue sea. Crenarchaeota, which make the temperature-adaptive enzyme, comprise a large fraction of total marine picoplankton, equivalent in cell numbers to bacteria at depths greater than 3,200 feet. The fraction of Crenarchaeota increased with depth, reaching 39% of total DNA-containing picoplankton detected, they reported in *Nature*. The average sum of archaea plus bacteria detected by 16s rRNA sequences ranged from 63 to 90% of total cell numbers at all depths. The high proportion of cells containing significant amounts of ribosomal RNA suggests that most microorganisms in the deep ocean are metabolically active. "Furthermore, our results suggest that the global oceans harbor approximately 1.3 × 10(28) archaeal cells, and 3.1 × 10(28) bacterial cells. Our data suggest that pelagic Crenarchaeota represent one of the ocean's single most abundant cell types," they wrote.

More work at identifying microbial life in the open ocean zones of the North Atlantic was conducted by scientists at the Department of Biological Oceanography and Department of Physical Oceanography, Royal Netherlands Institute for Sea Research, and the Max Planck Institute for Marine Microbiology. They performed gene expression profiles of water samples and concluded that "planktonic archaea are actively growing in the dark ocean, although at lower growth rates than bacteria and might play a significant role in the oceanic carbon cycle."

Open ocean archaea are primarily autotrophic, obtaining energy from inorganic physical or chemical sources and converting the energy into the organic components of their cells. Most bacteria in the open ocean are heterotrophic like us, relying upon organic molecules for energy and nutrients. They absorb soluble organic matter or engulf particles of food like a little Pac-Man. The consortiums of microbes cooperating on the *Titanic* include chemolithoautotrophs and heterotrophs, which give the *thing* access to energy from just about anything.

The bathypelagic zone where the archaea are most abundant extends from 3,200 feet down to a little over 13,000 feet. This is where giant squid roam. But no plants are found in the deepest regions. Most anything larger than a hair feeds on microbes or particles called marine snow. Bacteria and archaea also get nutrients from marine snow. Marine snow is a phenomenon scientists discovered with ROVs. The snowflakes consist primarily of decayed organic matter: the detritus of fish, plankton, and bacteria that are continually drifting to the bottom from the sunlit layer where there is much tearing of flesh by fish and grazing by zooplankton. There's also a lot of pooping taking place by everything from zooplankton to fish to ship passengers. Marine snow may be a barometer of the balance of life between phytoplankton, zooplankton, and fish. Less fish = more marine snowfall.

Susumu Honjo at Woods Hole described marine snowflakes as

"sticky and fibrous like a crumbled spider net." He found that the snow contains lots of fecal matter from zooplankton. The microbial feces are covered with goo that helps organic particles adhere and form aggregates, which sink as they attract other fecal pellets, dead foraminifera, airborne dust (containing iron) that settles into surface waters, and other heavier particles. As aggregates descend, they steadily accumulate more particles, becoming heavier and faster. Honjo found that aggregates often break apart, but the spilled particles are picked up or "scavenged" by other falling aggregates. Normally, a large portion of the organic matter in marine snow is eaten by fish in the mesopelagic zone, which is then ejected in their feces and recycled by microorganisms in upper and middle water columns that excrete the material again. This recycling is an extremely efficient use of nutrients before the marine snow falls to the abyss and deposits all that carbon that is supposed to be scrubbed from the atmosphere. The bottom sediment of the ocean is the "carbon sink."

According to Honjo:

> The removal of carbon from the ocean's euphotic (sunlit) layer to its interior carbon "sink" is critical to the process that keeps Earth's carbon cycle in order. We have learned that the speed of carbon settling to the ocean's interior is very rapid: Particles can travel from surface waters to the abyss in only a few days or weeks. Nature accomplishes this process ingeniously by wrapping labile organic carbon up in a package and ballasting it with calcium carbonate, which causes it to settle at high speed to the deep ocean environment.

Pellegrino wrote about marine snowfall increasing during the 1990s in the North Atlantic. He and Cullimore suggested it was due to overfishing of the North Atlantic's Outer Banks. The cod industry collapsed decades ago. Now fishermen are taking everything. If zooplankton's predators are declining, their populations will naturally explode. The zooplankton will overgraze the phytoplankton. The recycling of nutrients in the mid ocean will decrease. Marine snowfall will become richer in nutrients. The nutrients will become trapped in the sediment along with carbon. If overfishing is occurring worldwide, which it is, then the phenomenon of increased marine snowfall and increased trapping of nutrients in the ocean sediment will occur on a global scale, said Pellegrino and Cullimore.

Microbes will adapt to the changes just fine, but fewer phytoplankton and less efficient nutrient cycling in the world's oceans cannot be good for us.

Researcher Jay Quade takes soil samples in the Atacama Desert of Chile. Colonies of archaea were found thriving only a foot below the surface of the driest desert on Earth. The discovery has significant implications for life on Mars. *Courtesy of Julio Betancourt, U.S. Geological Survey.*

Photograph by *Spirit* shows the Martian surface appears similar to the Atacama Desert of Chile. New molecular probes are being designed to test for the presence of microbial life on Mars. *Courtesy of NASA.*

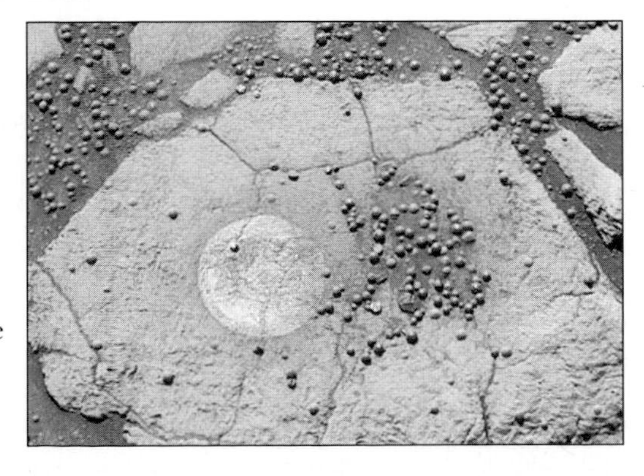

Unusual features found on Mars by *Opportunity.* These marble-size "rocks" indicate the former presence of water. Wherever water exists, life is a possibility. *Courtesy of NASA.*

Plate 2

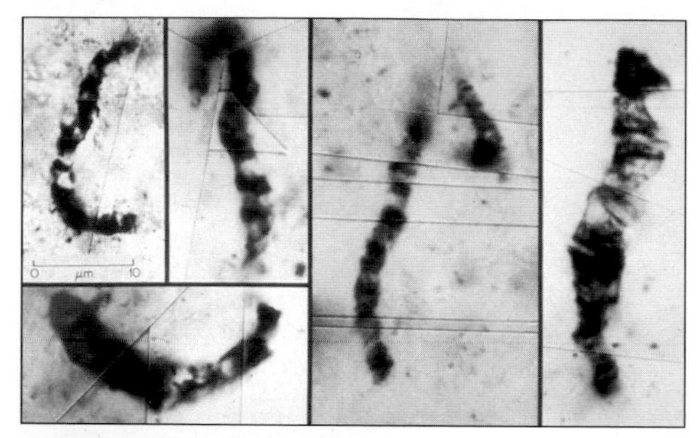

Oldest known microfossils, possibly from microbial mats. The microorganisms would have been anaerobic. Ancient microbial mats probably consisted of complex cooperatives of multiple-species microorganisms. *Courtesy of William Schopf.*

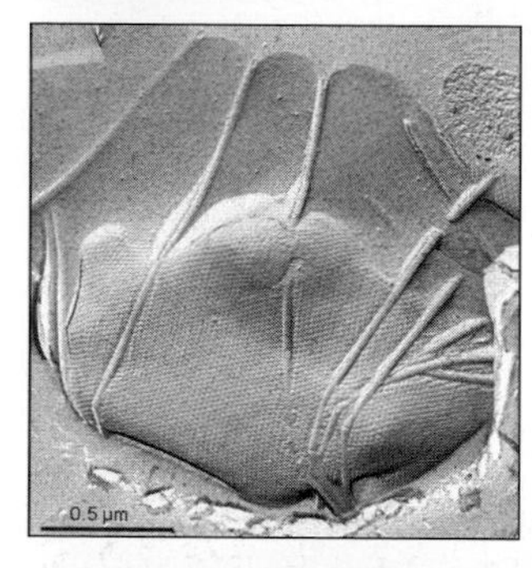

Pyrodictium, one of the hottest archaea alive, discovered in the 1980s by Karl Stetter. *Courtesy of Karl Stetter.*

Miriam Hernandez of INBio records the temperature of soil during sampling for termites in Costa Rica. Microbes living in termite hindguts have important implications for manufacturing biofuels. *Photograph by Hillary Theakston, courtesy of Diversa.*

Eric Mathur of Diversa collecting insects in a Costa Rican jungle from a light trap. Microbial DNA will be extracted from insects' intestines for biotechnology applications. *Courtesy of Diversa.*

Photograph of Carl Woese. *Photograph by Don Hamerman ©2007.*

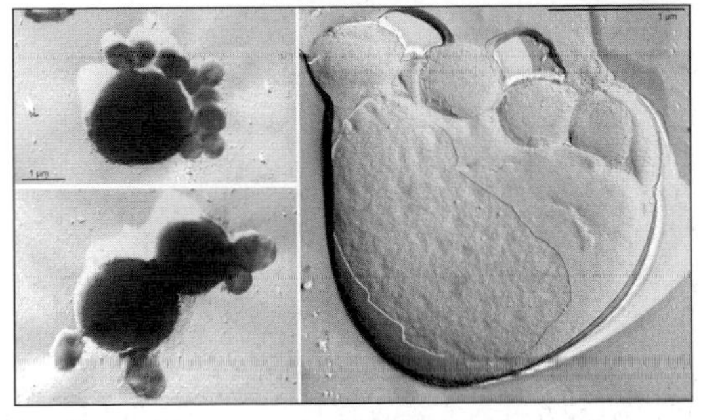

Smaller spheres are parasitic *Nanoarchaeum* living off the metabolism of its host *Ignicoccus. Nanoarchaeum* possess genes for DNA repair and reproduction but lack genes for their own metabolism, giving them a wide variability in selection of hosts and environments. Such parasitic organisms could survive in outer space. *Courtesy of Karl Stetter.*

Archaea are grown in vast quantities of ceramic-lined titanium vats under extremely high pressures at near boiling temperatures. The incubators were custom designed by Karl Stetter to simulate conditions thousands of feet deep at hydrothermal vents on the ocean floor. *Courtesy of Karl Stetter.*

Plate 4

The *Mir I* Russian submersible is being launched for a 10-hour round-trip dive to the *Titanic*, which is covered in one of the most complex consortia of microorganisms on Earth. *Photograph by Tim Friend.*

A close-up of a rusticle, which is a consortium of microorganisms harvesting iron from the *Titanic*. *Photograph by Lori Johnston.*

Formations of living microorganisms harvesting iron from the *Titanic*. These formations creep slowly along the ship over time. *Photograph by Tim Friend.*

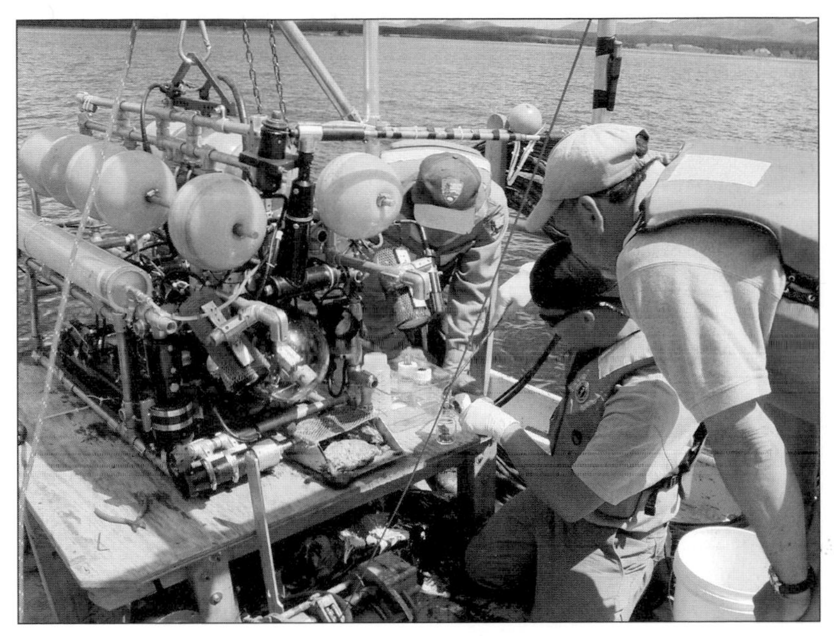

A National Park Service biologist and remotely operated vehicle specialist Dave Lovalvo retrieve sediment samples containing new species of archaea from Yellowstone Lake. Lovalvo built the $500,000 underwater robot, which is outfitted with sampling equipment, sonar, and a video camera. *Courtesy of Leif Christoffersen.*

Slide depicting spires. *Photograph by Dave Lovalvo.*

Jay Short holds a piece of an ancient spire constructed in part by microbes as long as 10,000 years ago. *Photograph by Leif Christoffersen.*

Plate 6

The Valley of the Geysers. Hydrogen sulfide gas escapes from fumaroles, which line the valleys. *Photograph by Ryan Short.*

Boiling mud contains thriving colonies of archaea. *Photograph by Ryan Short.*

Karl Stetter sampling for archaea at a hot spring at the Valley of the Geysers in Kamchatka. The region may resemble very early conditions on Earth. *Photograph by Ryan Short.*

Professor Simon A. Wilde of Curtin University examines in 2001 the outcrop of the Jack Hills metaconglomerate where he and others discovered detrital zircons that formed as long as 4.4 billion years ago. These are the oldest known samples of Earth. Geochemical analysis led to the hypothesis of a Cool Early Earth that was possibly habitable, rather than the inhospitable Hadean conditions envisioned earlier. *Courtesy of John Valley, University of Wisconsin.*

Plate 8

The surface of a microbial mat collected from Area 4 in the Exportadora de Sal saltern system in Guerrero Negro, Baja California Sur, Mexico. The complexity of the surface can be seen. The white areas are microbial populations that appear to occupy the highest spots on the mat. *Courtesy of NASA.*

Fossilized microbial mats found in Western Australia dating back 3.5 billion years. These unusual formations stretch over a six-mile area. Much of Earth's surface could have been covered by vast cooperatives of microorganisms. *Courtesy of Abigail Allwood, Australian Centre for Astrobiology.*

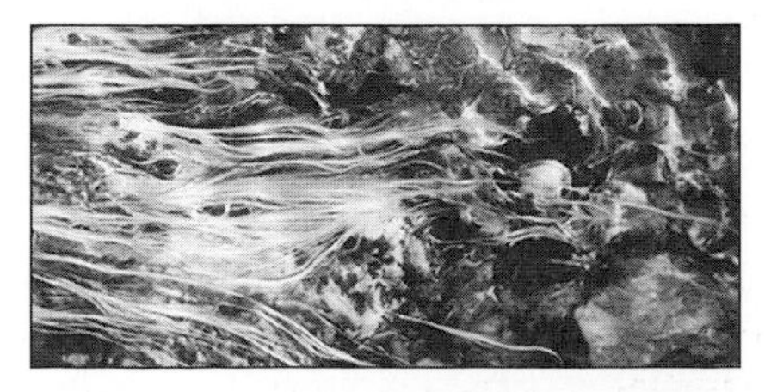

This type of microbial mat contains streamers of *Aquificales* that are attached to thicker layers of microbes beneath. *Courtesy of NASA.*

After an hour of lying prone and reaching a depth of about 6,000 feet, I removed the jacket from my head and sat up to join Anatoly and Harris. We had at least an hour before reaching bottom. I asked Anatoly whether he had observed an increase in marine snowfall in the 10 years he had been regularly diving on the *Titanic*.

"Of course," he replied matter-of-factly. "It is because of the fishing."

Marine biologists from the Shirshov Institute have documented an increase in zooplankton populations. Indeed, during my expedition, Russian scientists were casting nets and fine seines as deep as 5,000 feet to sample both fish and zooplankton populations from the mid ocean. I spent most nights out on the stern deck helping them collect samples into jars for study in the ship's laboratories. With marine snow carrying more nutrients, population growth might boom among the bottom-feeders and microbes. Anatoly said the heavy marine snowfall was feeding the rusticles and accelerating their growth. This in turn was transforming the *Titanic* into an oasis for species that do not normally live at 12,600 feet. Crab populations are rising, documented by videos taken by treasure hunters and by robotic cameras sent into the ship's corridors and interior rooms. The iron-loving *thing* has turned the *Titanic* into an oxygen generator. Oxygen levels around the *Titanic* are about 6 parts per million (ppm), the same as tap water for your fish bowl.

The average depth of the ocean in the abyssal zone is about 13,000 feet. O_2 levels should be around 1 ppm or less. Water temperatures are 33 degrees F, and pressure is at least 6,000 pounds per square inch, enough to immediately disintegrate us. A pinpoint spray at this pressure would behead you. Any implosion would turn us into an organic garden for heterotrophic microorganisms, while the lithotrophs would have a go at the submersible's metals, working their way through the wiring, ballast tanks, and finally the nickel-steel hull.

Anatoly twisted in his seat and opened a compartment behind him. It was a treasure chest loaded with sandwiches made with Wonder Bread, mayo, cold cuts, iceberg lettuce, and cheese. We each got a bag of chips and sodas or bottled water.

As I ate, Anatoly and Harris discussed the possible location of the diamonds. Anatoly wasn't buying the diamond legend at all, but he allowed if the diamonds exist, they're deep inside the collapsed decks of the interior and impossible to reach. Anatoly suggested maybe in another hundred years or two, when the *thing* has finished extracting all of the iron from the hull, deck floors, and vintage automobiles in the cargo bays, a new generation of treasure hunters will discover it sitting atop a slag of iron ore.

According to Cullimore's calculations, the *Titanic* should be reduced to a layer of iron ore as early as 2106 with only the massive engines still intact and exposed. He predicts it will take the microbes another 5,000 to 15,000 years to finish off the hardest steel and nickel. (See Plate 4.)

Anatoly stopped talking to Harris and radioed topside. The Russian-speaking topside commander sounded even more exotic with the static and reverb created by two and one-half miles of water between *Mir I* and the *Keldysh.* Anatoly glanced over with a smile. I sensed that we were no longer moving. Anatoly informed us that we had reached the bottom. Harris and I scrambled into prone positions with our noses pressed against the view ports. Anatoly switched on the powerful 5,000-watt high-intensity lights attached to the front of the sub, illuminating a nearly barren desert as Anatoly steered northward. The sediment was pure as the white sands of New Mexico, sparsely dotted with single-stalk coral and white sponges that looked like dandelions. We glided slowly about 10 feet above the sediment to prevent the thrusters from stirring it into a thick cloud. The topmost layer was all marine snow.

The ocean at this depth contains very little dissolved oxygen, which turns the abyssal zone into a cold-storage locker for nutrients. Every few thousand years the world's oceans completely exchange water top to bottom, tilling nutrient rich sediments to enrich the depleted epipelagic zone. Unless we're on our way to becoming Venus, all the nutrients being trapped in the current warming cycle will one day stimulate a new and abundant cycle of ocean life. Meanwhile, to keep the pot stirred, localized upwelling is always stirring nutrients—normally, anyway—in some region or other depending on currents and water temperatures. One of the problems with climate warming is it generates sharper temperature gradients between the sunlit layer and the cooler mesopelagic zone, thus inhibiting even the localized mixing of nutrients. I'm not sure we realize how closely microorganisms, zooplankton, phytoplankton, fish, and climate are tied to each other and to the well-being of life as we know it. I'm positive that most of us do not appreciate that microbes are quite likely the most powerful factor in the whole equation.

The hadal zone is made up of the ocean's deepest trenches, the record holder being the Marianas Trench at 35,798 feet. We're only as deep as the *Titanic*, which lies on the abyssal plain at 12,500 feet. As we glided closer to the ship, an occasional rattail fish with its huge head, bug eyes, and long tail slithered effortlessly along the bottom. Giant purple sea cucumbers shined like jewels on a white desert. Scattered bits of metal debris appeared through the porthole, the first signs of the wreckage to come. After a few more silent minutes, Anatoly slowed the submersible to a dead stop. With a toggle of the thrusters, we tilted upward and rose nose first.

"Now, *Titanic*," Anatoly said quietly.

Through the greenish haze of the floodlights, the iconic bow came into view. It was difficult to grasp that this image was real, that the *Titanic* was only 50 feet away on the other side of the view port. Had I been alone I would have wept.

The ship's massive anchors were intact and completely covered with long rusticles, just as Cullimore and Pellegrino promised. As we moved slowly upward, the bow railing appeared, rust icicles dangling as Ballard first described them, but longer. No one in the sub spoke as we hovered below the bow making a slow arc from the port side to the starboard. We glided slowly upward until *Mir I* was straight off the bow. I began shooting photos through the view port with the digital camera. Our sub passed over the bow railing, revealing the deck with the capstans and another massive anchor. The gaping mouth of a cargo hold beckoned, but we dared not take the sub inside. We flew to the upper deck behind the bridge, which appeared as fragile as wet cardboard. The telemotor, which held the ship's steam-assisted wheel, was intact. At its base was Ballard's original memorial plaque, joined by a slew of others left by Tulloch's expeditions.

The walls of passenger and crew cabins, including the captain's quarters, which Anatoly said were intact a year ago, have fallen away. Ceilings have collapsed along the upper decks. We hovered next to Captain James Smith's quarters. His white porcelain claw-footed bathtub, only partly visible a year before, was fully exposed.

"This is terrible," said Anatoly, who has spent more than 200 hours at the *Titanic*. "I have never seen her in this bad condition."

Much of the ship's outer surface was enveloped with orange bumpy "rust." Occasionally a section of deck or hull appeared clean, but Cullimore said that meant the entity was extracting iron from the other side of the steel. The super-organism had become so monstrous that it was engulfing the entire ship. The wreck displayed at least three basic types of rusticles—the dangling icicles; tight concentric clumps that appear as leeches on the hull; and a third that spreads like a thin veneer, its outer layer protecting the consortiums of microbes, its inner layer penetrating gaps and cracks in the steel plating. Cullimore and Pellegrino found rusticles embedded with coal, glass fragments, clays, and stones—ejecta from the *Titanic*'s impact with the ocean floor. But the living *thing* is dominated by iron-loving microbes, sulfate-reducers, and fungi. New species are joining the monster as it grows.

The grand staircase, immortalized in photographs, was a dark hole. The staircase had vanished by the time expeditioners snapped pictures from *Mir I* in 1991. The surrounding columns, which were visible then, were gone. Large sections of the ship's superstructure were separating. The mast had fallen across the bow and extended to the forecastle deck, ready to cave through portions of the forward deck.

We flew to the port side. Anatoly gingerly positioned *Mir I* next to the Marconi radio room where a crew member sent the *Titanic*'s final distress signals. The wireless radio-telegraph lies beneath the rotting metal deck just beyond the room's former skylight. Harris wanted to remove the deck above the radio room and retrieve the telegraph before it caved through the floor and vanished. He stressed to Anatoly the significant historical value of retrieving the radio. Anatoly said nothing.

We spent two hours surveying the wreck and examining rusticles, then glided toward the end of the bow section where it ripped from the stern at 2:17 a.m. on Monday morning, April 15, 1912. Three minutes later it plunged to the bottom like an asteroid. What carnage of steel, coated in reddish orange, yellows, and browns. Anatoly used the manipulator arm to touch a portion of a rusticle that had taken over a hulk of steel torn from the ship. Immediately a cloud of iron-rich fluid gushed out from the tendril. It looked like blood. The *thing* had even developed an immune system, Pellegrino said, manufacturing at least a half dozen different antibiotics.

Cullimore called the massive patches and the dangling rusticles "consorms," short for consortial forms. Clinging like gargantuan leeches to the wreck, their vascular systems have a minimum internal surface area of 6,280 square miles. The mass is estimated at a minimum of 650 tons and includes up to 175 tons of iron extracted by the microbes from the ship's steel. The *thing* harvests 20 five-ton truckloads of iron from the ship per year. At the current rate of microbial activity, the main deck could collapse into the hull by 2028. Henrietta Mann has suggested that in addition to iron and sulfur, microbes on the outer layers of the rusticles are metabolizing organic nutrients from marine snow and channeling metabolites into the vasculature.

Beyond the twisted bow wreckage was the debris field. It is the last testament to the *Titanic*'s human tragedy. Boots, still laced, lie there. Plain bone china reflected the intense light from the sub. Cups, dishes, and wash basins used by passengers from mostly second- and third-class compartments were scattered across the bottom. Most of the first-class relics had already been taken. We passed by several bowler hats and tried to grab one. It ripped at the touch of the mechanical arm. Pellegrino wrote about ghost-like remains of metal objects, digested by microbes from the inside. A metal window frame that appeared brand new disintegrated when we touched it with the sub's robot arm. It was merely a carbon shadow. The cold temperature, intense pressure, and near absence of oxygen around the ship hold images of these objects intact though they are really no more than dust.

Nothing much should be living on the abyssal plain except for anaerobic microorganisms colonizing the sediments. But the rusticles are thriving beyond comprehension. At hydrothermal vents, microbes obtain power from spontaneous chemical reactions in the superheated fluids. But nothing hydrothermal was happening there. I began wondering what could be the source of power for the *Titanic*'s microbial megalopolis. The answer was simple and very clever. Place two or more dissimilar types of metal near each other in seawater and they will generate an electrical current. Ships are made of dissimilar metals such as brass, different grades of steel, nickel, and aluminum. Parts of the wreck are negatively charged while others are positively charged. The *Titanic* was humming with electrical currents and the microbes were exploiting them to power their industry.

"The whole animal kingdom is driven by electricity actually. But when it comes to the microbes that are in the rusticles, they are generating electricity the whole time and that is part of why the iron is coming out of the steel. When you make a metal you embody a lot of energy into that metal. When iron erodes out of steel it also naturally releases some of that energy through oxidative reactions. If you have dissimilar metals they automatically exchange electrons. This is well known as electrolysis. But what is not well known is when you put microbes into the dissimilar metals they have a ball," Cullimore said.

Cullimore and Pellegrino wrote a report in the late 1990s after discovering that the *Titanic*'s iron-metabolizing microbes "have entered into a symbiotic linkage with other microorganisms. The form and function of the rusticles is now under intense scrutiny because they appear to be a new doorway through which we can see, at a distance, our own multicellular origins. They are not a single species of plant or an animal but a complex polymorphic (many shaped) consortium dominated by microorganisms, each performing some specific but essential function for the ongoing functioning of the rusticle."

From Cullimore and Pellegrino's perspective, the *Titanic* hull plates are the "dinner table" for rusticles and "iron was the main course." The *Titanic*'s steel was made from strips of hot iron rolled and pressed producing very fine layers into which the microbes are able to insert themselves and form colonies. "After entry into the steel plating, the microbes cleave the steel like a biological wedge, forcing the layers apart at each cleavage, essentially doubling the surface area available for growth. Many of the rusticles that were 'washed' from the hull plates had a thick skin of metal (up to 0.12″ thick) still attached to their bases, and their interiors often contained sheets from other layers, displaying a fractured series of biological wedges. As the rusticles penetrated into the steel, there occurred a gradual increase in the ratio of surface area relative to volume (via microbially induced cleavage and cavity growth). This is probably a major cause for the acceleration of rusticle growth (and the associated deterioration of the *Titanic* itself) observed by repeat dives to the site since the first robotic reconnaissance in 1985," Cullimore and Pelligrino wrote. "Identifying which microbes are involved and theorizing about how they manage to 'digest' steel is not the stuff of passion. The real excitement here is in the unexpected door that has been thrown wide open, providing a view of the deep wilderness that reaches far beyond the *Titanic*—perhaps 3 billion years beyond to the first consortial life forms, things that became us."

After Cullimore's first expedition he began growing rusticles in his laboratory on steel ingots, watching them form complex attachments. The growths showed how consortiums develop and gave him ideas about controlling the corrosion of machinery in marine environments, such as steel oil rigs, commercial ships, water and oil pipelines, and his water wells.

Cullimore and Pellegrino stated in their 1997 report:

> Essentially, a new branch of science is being born. During the first decade after the discovery of the Titanic, robot telepresence and submarine "fly-bys" with video camera logging had led most observers to view the rusticles simply as nuisance mineral formations of little note (rather like the stalactites in caves). When rusticles were finally recovered, observed, dissected, analyzed and cultured in 1996, a new, remarkably durable and robust form of life revealed itself at a site where most humans would think life could hardly exist let alone thrive. And strangest of all is to peer into the heart of a rusticle, and to recall that for eleven years, in virtually every photograph of the wreck site, these "creatures" were staring us straight in the nose. **Each rusticle is a complex assemblage of structures displaying many characteristics normally associated with tissue differentiation in higher organisms.** Each displays a series of microbial consortia dominated by bacteria and fungi—which are, in essence, independent free-living cells grouped into differentiable structures. Included among these structures are internal water channels, water reservoirs, hardened iron-rich plates, surface ducts which pass through the external hardened plates and connect the rusticle interior to the outside environment, **porous sponge- or pumice-like layers, bundled fibrillar clumps and elegant thread-like "girders"**, which appear to be strung through the structures and channels and apparently provide a measure of mechanical stability to the entire rusticles.

I highlighted the above in bold because rusticles are not the only biogenic structures found with similar architecture. This is a common pattern, and a very ancient one at that. Pellegrino suggested the metal "tissue" of the rusticle is the primitive beginnings of something akin to animals and plants. It seems that life on Earth is genetically programmed at the most basic microbial level to organize a vasculature. By definition this is an arrangement of blood vessels in a body or an organ. But we also see the pattern in plants.

The hanging rusticles of the *Titanic* developed a central water channel that connects to reservoirs surrounded by sac-like extensions that connect to smaller channels that form capillaries throughout the structure and connect to ducts on the surface. Cullimore and Pellegrino developed two theories about how fluids are channeled through the primitive circulatory system. One theory is that pressure builds when carbon dioxide dissolves and forces water through the channels like a fizzing soda. Cullimore called this a "pneumovective" process because it involves changes in the surface tensions that occur as water bubbles through the channels. Another theory is gravity. As iron dissolves out of the steel into the water, the fluid becomes heavy and descends through the circulatory system. Microbes absorb iron from the water, metabolize the iron, and respire iron ore, which becomes incorporated into the outer layers of the rusticles like microbial armor. After microbes remove iron from the water, the water becomes lighter and either circulates upward or is continually ejected through ducts at the bottoms of the rusticles as heavy water descends.

"One feature common to all rusticles is a system of bacterially-derived threads, which line or crosshatch the channels, plates and reservoirs. They appear to bind the whole structure together into something that can be thought of as a 'woven sock of slime.' Some of the threads are very durable. They survived the partial collapse of the rusticle structure, dissection, microscopic examination, transportation, drying and storage. . . . Clearly iron is being harvested from the *Titanic* by the biomass, sometimes referred to affectionately as 'the Rusticle Park.' . . . As deep ocean microbes convert the *Titanic*'s mass to their own, it becomes possible to say that something is indeed coming to life on the *Titanic*; and it is the *Titanic* itself."

Moments before the *Titanic* plunged to the bottom this ocean area was an empty abyssal plain, the sediments filled with subsistence farmers making a living from microscopic particles of iron, sulfur, phosphates, and other essential nutrients. Look at what they created in less than a century. I came to see the ship, but instead witnessed an unimaginable life form with a metallic, flesh-like body. It was the most wonderful *thing* I have ever seen. The *Titanic* is alive, and it has become something more than colonies of microorganisms. The question is what?

After four hours on the bottom Anatoly informed us it was time to return to the surface. During our ascent I felt haunted, not by the unfortunate souls, but by what the ship has become. What is programmed into the genes of microorganisms that can create such complexity and efficiency on a monstrous scale?

At the surface, the weather had turned bad. Recovery of the submersible took more than an hour. We rocked violently on the surface. The danger of retrieving a submersible in bad weather is primarily to the people who must climb onto the sub to recouple the cable. These people could be crushed between the sub and the ship's hull or be lost in high swells and swept away by the current. If not, the sub could slam against the hull as the crane hoists it from the sea on a yawing ship.

A *Keldysh* tradition after diving is enjoying Stolichnaya vodka, caviar, smoked salmon, black bread, and a few beers with Anatoly in his quarters. He and his wife live on the ship for much of the year. They have autographed *Titanic* documentary and *Titanic* movie posters framed on the walls next to beautiful Russian paintings. After boisterous conversation buoyed by adrenaline and vodka, we took a sauna and intermittently cooled down in a 40 degrees F splash pool fed by the North Atlantic. It turned into a rather late evening. The experience of seeing the *Titanic* and the rusticles left me dumbstruck the next day. I feared speaking about it, lest I break the spell. The deaths of passengers began sinking in more than I anticipated. The story of their final moments is told in hats, gloves, shoes, and luggage packed for the voyage to New York, all strewn along the ocean floor. The *Titanic* is a memorial. It should be left alone.

Seven years have passed since my expedition to the *Titanic.* The haunting memories of the passengers' belongings have faded, but that consortium, thriving in the abyss unnerves me still. What is it about the *thing* that creates such a disturbing feeling in the pit of my stomach? I'm guessing it is the eerie similarity of the metallic, microbial slime to our own tissues.

Few scientists besides Cullimore, Pellegrino, and Lori Johnston have studied the rusticles in any depth. Mann, now at Dalhousie University in Halifax, keeps peering into the rust and finding more organisms. She and Bhavleen Kaur announced in August 2006 their discovery of a new species of bacteria in a rusticle. This one aggressively corrodes metal. It belongs to the genus *Halomonas*—the salt-lovers. More recently, cousins of the corrosive, iron-loving "BH1" have been found in the chilly waters of the Antarctic and on low-temperature hydrothermal vents. BH1 appears to be an ordinary cold-water extremophile recruited into an extraordinary consortium, becoming one of the most productive members of the *Titanic*'s new society.

The *Titanic* holds more secrets that anyone dreamed possible. It could shine a new light on the social order of microorganisms and the tremendous metabolic efficiency of consortiums. Consortiums are where microbial ecologists may find answers about how life crossed the threshold from simple colonies to complex organisms. Somewhere in the genes of these microorganisms lies the blueprint of higher organization. Maybe the rust that never sleeps will awaken us to a new view of life.

6
US . . . AND THEM

We used to think that if we knew one, we knew two, because one and one are two. We are finding that we must learn a great deal more about "and."
Sir Arthur Stanley Eddington

"So what are you wearing?" I asked after being awakened at 1 a.m. by the phone ringing.

"A five millimeter wetsuit, hood, heavy gloves, and thick booties" Eric Mathur said at the other end of the line. He paused, uncertain. "What do you think?"

It was 10 p.m. in San Diego. Mathur couldn't wait until the next morning to call with the news.

"What did you say the water temperature will be?" I asked, still waking up. I switched on a small lamp next to the bed and sat up, bare feet on the floor.

"Between 38 and 41, 42 degrees. Maybe a seven is better?" Mathur paused again, then with his endearing exuberance he nearly shouted, "This is going to be so wild, man. Only one other team has ever been diving there. Ever. I told you we'd make this happen."

He most certainly did. Mathur had called Jay Short the same night he and I watched with awe the underwater video of the freshwater hydrothermal vents back at the October meeting of the Thermal Biology Institute at Yellowstone National Park. Short immediately saw potential and approved the idea. Mathur began plotting with John Varley, Yellowstone's chief scientist, the very next morning. They put together a proposal during the ensuing months and the Gordon and Betty Moore Foundation awarded them a $132,500 grant. Within a few days, scientists from Yellowstone National Park, led by Varley, and a team from Diversa, headed by Short, collected samples from hydrothermal vent systems on the bottom of Yellowstone Lake to "cultivate the uncultured" using Diversa's automated microbial ecology method. The grant was for a pilot study called the Molecular All Taxa Biodiversity Inventory (MATBI).[1] The project tested how well Diversa's high-throughput culturing method can identify and characterize the invisible inhabitants of the water and sediments of Yellowstone Lake's hydrothermal vents. Depending on its success, Yellowstone scientists will use the data to build up their microbial taxonomy of the lake, which, through no fault other than the limitations of technology, hasn't amounted to much over the past 137 years of the park's history.

Yellowstone Lake is the highest big lake in North America at an elevation of 7,732 feet. It has a surface area of 136 square miles, is 20 miles long, 14 miles wide, and at its deepest officially 402 feet. Various groups using sonar claim to have found a hole 600 feet deep. It is the second largest high-altitude lake in the world. The average water temperature is 35 degrees. The surface of the lake stays frozen with three to four feet of ice half of the year, solid enough to drive a Mack truck across it in most areas. By July 2004, only 263 species, mostly algae, protozoa, plants, and fish, had been identified in Yellowstone Lake. Only a few of the species represented bacteria and no archaea species had been published. Scientists assumed the lake had a low level of life, especially sparse in microorganisms.

The technology Diversa used for the MATBI was the same it applied to the thousands of environmental samples taken during their many worldwide expeditions. Diversa had been honing the method for several years. After securing their patents, they disclosed it to the scientific community in October 2002 in the *Proceedings of the National Academy of Sciences.*

Remember that the biggest evolution in microbial ecology, which Norman Pace contributed to in the mid-1980s, and Jay Short expanded dramatically in the 1990s, was breaking from the restraints of having to grow organisms in culture to identify them. A full century after Koch invited the first microbes to a rather ordinary dinner of agar, scientists still could culture only 1% of organisms present in an environmental sample—the so-called microbial weeds. Pace had adapted Carl Woese's painstaking method of sequencing the 16s rRNA genes of microbes by hand and combined it with the polymerase chain reaction (PCR), giving microbiologists the freedom to finally explore the microbial diversity of any environment independent of culture and based solely on sequence data.[2] Short expanded the capability even farther. The newer methods finally uncloaked the invisible microorganisms that scientists were certain were hiding under their noses all along. Scientists knew that many more species had to exist, but the scale of what they ultimately found was vastly underestimated.

Metabolically, Pace's method was like holding up a mirror to detect the fog of breath from a ghost. The presence of previously unidentified gene sequences in a sample of soil or water reveals biological activity. Armed with sequence data, microbiologists could do additional detective work to isolate individual microorganisms and attempt to culture them, Stetter-style, to learn about their individual metabolisms and personal habits. This isolation and culturing process, however, continues to be painstaking for most labs with limited resources, and it still involves about as much art as science.

In 2003, Craig Venter took environmental sequencing to the next level with his quest to survey the microbial diversity of the planet's oceans aboard the 90-foot marvel of a sailing yacht, *Sorcerer II.* Venter and his crew began casting seines fine enough to capture viruses and picoplankton in the seas from Cape Cod to Cape Horn to Bali and beyond. The expedition made the cover of *Wired* magazine and generated newspaper headlines worldwide. Scientists at the Venter Institute were extracting DNA from samples shipped back for shotgun sequencing, generating unprecedented volumes of sequence data associated with previously unidentified microbes. Venter smartly applied to microbial ecology everything he learned from sequencing whole genomes of microorganisms and from sequencing the human genome, his own.

The only drawback with sequence data is that they are data. Scientists who want to learn the most about individual microbes must eventually isolate them and observe them directly through cultures.

Short's method takes us directly from sampling to pure culturing. It involves capturing individual living microorganisms from environmental material, encapsulating the individuals in micro-droplets of a gel, and then conducting what Short calls "massive parallel microbial cultivation" using nutrients selected from the natural communal environments of multiple species. The colonies of living microbes provide direct access to metabolisms, behaviors, and the biochemical relationships between members of different species from the same neighborhood as well as geochemical relationships between species and their environment. This is an incredible achievement.

All of the microorganisms that have resisted culturing efforts for the past century can swim, squirm, and squiggle, but no longer can they hide from the ingenious technique developed by Short and his team of investigators—Martin Keller, Karsten Zengler, Gerardo Toledo, James Elkins, and Eric Mathur of Diversa, and Michael Rappé of Oregon State University. These scientists have closed a loop begun more than four centuries ago when microscopes opened the first portal into the invisible universe. What Short and company have done is exactly like setting up 10 million microscopes, each holding an individual organism on a slide, and simultaneously cultivating all 10 million organisms with a single burst into microbial hyperspace.

As Mathur said, "It's totally radical."

For Diversa, Short's massive parallel microbial cultivation is merely the first warp drive of the greater biotech engine he designed for developing commercial enzymes from raw DNA. But for microbial ecology it is by itself an astonishing advance that perfectly compliments the Venter/Department of Energy (DOE) brand of supersonic sequencing. After Diversa cultures all of the microorganisms from the Yellowstone samples, the DOE's Joint Genome Institute (JGI) will sequence the microbes. The end product for the MATBI will be cultures of virtually 100% of the representatives of microbial populations present in the samples *plus* the genetic sequence data needed to construct phylogenetic trees of all three domains of life based on the 16s rRNA genes of archaea and bacteria and the 18s rRNA genes of eukaryotes. The MATBI is powerful and unprecedented microbial ecology, as good as it gets in the present.

The Moore Foundation and Yellowstone National Park saw the MATBI as an extremely valuable opportunity for the microbial ecology of the lake. JGI agreed and decided to donate its sequencing power to the project. The MATBI was unique in that it was a collaboration between the National Park Service, the DOE, and a for-profit biotechnology company. The MATBI also represented the first complete cultivation and characterization of an entire community of microorganisms on a hydrothermal vent system and its surrounding waters. The fact that Yellowstone's is a rare freshwater vent system made it even more significant. Jay saw the MATBI as an opportunity for Diversa to finally demonstrate its technological power.

Only a few lakes on Earth are known to possess freshwater hydrothermal vents. In the late 1980s, scientists found sulfide deposits in Lake Tanganyika, East Africa, that suggested the presence of vents. In 1991, scientists discovered volcanic vents on the deep bottom of 25-million-year-old Lake Baikal in Eastern Siberia. Crater Lake in Oregon has a hydrothermal vent system. A few high-altitude lakes in the volcanic regions of the Andes have been found too.

But Yellowstone Lake offers the best access and opportunity to explore both extremes of temperature boundaries in the outer limits. The water is oh so cold, but down on the vents it is hot as blue blazes, as high as 248 degrees F in at least one spot that has been measured. Many of Yellowstone Lake's vents follow fault lines detected by scientists from the U.S. Geological Survey. The lake bottom in an area known as West Thumb is an open "vent field" with new ones appearing in recent years and even the past few months at this writing. Mathur said the West Thumb vent field was where we would be diving.

"So what are you wearing?" Mathur asked.

"You said 38 degrees?" I yawned, still waking up.

"Yeah. Killer, huh?" Mathur replied.

"There's no way I'm getting in that lake with a five-mil suit," I told Mathur. "I'd guess seven at minimum. But I'll ask somebody who does a lot of cold water stuff. I wouldn't miss this for the world no matter how cold it is."

We chatted a little more about the game plan for the dive and hung up. It was 1:30 a.m., and then I was wide awake. I dread cold water. (As a kid I would a wait a day before venturing into the inflatable pool my mom filled in the summertime with water from a garden hose.) The next morning I found a company in Huntington Beach, California, that makes a cold water suit for Navy SEALs. This lifesaver is the amazing Xcel Polar Tri-density Semi-Drysuit—one piece, hooded, NINE millimeters thick, insulated with titanium fibers. Only the tiniest amount of cold water seeps in thanks to a special neck dam and tight seals at the sleeves and ankles. The Xcel zips across your chest from shoulder to shoulder. I called the manufacturer right away, and they took my measurements over the phone. Six hundred dollars and three days later I was equipped for anything from a freshly filled wading pool to the Arctic Ocean.

Short called the same day I purchased "Gumby" and explained rather breathlessly that Varley wanted us to add a new component to the microbial ecology study of Yellowstone Lake's hydrothermal vent systems. Varley wanted us to explore an unusual underwater feature in the lake called the Bridge Bay Spires. Better yet, he wanted us to collect extremely rare samples of the spires. This was a fortuitous twist in a private obsession I had developed since August 2000 to find cousins, if they existed, of the *thing*. The spires resemble a cluster of ancient, eroding columns, some of them nearly 40 feet tall. Scientists generally assume that spires are mineral deposits, remnants of geothermal chimneys similar to black and white smokers at ocean bottom vents. But Varley, who looked at one sample of a spire from a previous expedition in 1997, had a different theory. He wanted to see if he was right. It turned out a few others who studied the spires more closely also had ideas about their origin.

Until then, I had found very little information about rusticle-like structures aside from articles by Cullimore, Pellegrino, Johnston, and Henrietta Mann. Sometimes this is a warning sign that the evidence does not exist or that the science is hokey. Not so in this case. The cousins of the *thing* were out there. It's just that the scientists who were studying them had not crossed paths. I think the scientists might also have been a bit shy about openly discussing what they were finding. In my case, after one too many raised eyebrows, I had stopped talking to microbiologists about the rusticles as a monstrous organism at an evolutionary threshold. But I had not stopped thinking about it. One can't after one sees it.

My preparation for the Yellowstone expedition initially led me to work published by scientists at the University of Wisconsin-Milwaukee, who had conducted more than a decade of pioneering research on Yellowstone Lake's hydrothermal vent systems.[3] The team had studied the lake's geothermal features, vent systems, and geochemistry. Some of their work fit like a missing piece in a puzzle with regard to the relationship between microorganisms and geology. Everything hinted at life's origins.

"Successful sampling of microbiota from shallow, freshwater hydrothermal systems with high temperature and elevated, sometimes toxic mineral content will provide an additional model for understanding biogenesis and evolution under primitive Earth conditions," Carmen Aguilar of the Wisconsin team told science writer Kelly Kizer Whitt about her team's work on the spires and vents. "Many of the appealing features of two primary theories of early life—that of hot, reducing conditions and shallow, atmospherically influenced pools—are melded in the Yellowstone hydrothermal vent ecosystem."

Yellowstone is a genuine wonderland. Models for the violent volcanic realm from which archaea might have sprung 4 billion years ago and for Darwin's warm little pond of chemicals possibly sparked to life by a thunderbolt all are neatly packaged in a single body of water and wrapped with 110 miles of shoreline. Yellowstone Lake's hydrothermal systems and frozen conditions six months of the year also represent a perfect model for studying the possibility of life on Jupiter's moon, Europa.

Wisconsin's James Maki, lead author of a paper titled "Investigating the Microbial Ecology of Yellowstone Lake," summed up microbial ecology about as well as anyone: "Microbial ecology is the study of microorganisms in relation to their biotic and abiotic environment. In practice, it has been described in a graduate student motto as 'the study of physiology under the worst possible conditions.'" Thomas Brock published that description in his classic 1966 textbook *Principles of Microbial Ecology.*

"More recently, microbial ecology has also been indicated to be the link between all branches of microbiology. In any case, similar to traditional ecology, microbial ecologists study individual organisms, populations of individuals, communities of populations, and ecosystems," Maki wrote. "This is done this with a variety of approaches and tools, including microscopy, culturing, molecular biology, and biochemistry. Much of what is studied by microbial ecologists revolves around three questions: Who is out there? How many are there? What are they doing?"

According to Maki's colleague Charles Remsen, Yellowstone Lake had been regarded over the decades as relatively sparse in terms of microbial life. Maki wrote that scientists assumed the lake had a "low metabolism and not much in the way of nutrients."

According to Varley, most of the water in the 140 tributaries flowing into the lake originates as snow, which would be very dilute in nutrients, close to distilled water. On its journey to the lake, the snowmelt flows mostly over rhyalite rock, which itself is poorly productive biologically and not very soluble. The calculated average residence time of tributary water in the lake is 11 years, so the nutrients that the streams bring into the lake are the sum of 11 years of input. The nutrients are concentrated over time by evaporation, and become trapped in sediments as the relatively few known species of fish, plankton, and other organisms die. "The result," Varley said, "is the lake water appears to be low in nutrients and not conducive to an active universe of microbes."

The Wisconsin team began working at Yellowstone Lake at least since 1985. As they studied the lake more closely they found higher levels of nutrients than expected for the presumably low population of microbes. This suggested that the lake was more metabolically active than it appeared—perhaps a case of invisible microbes. Time to bring out the mirror and do a little ghost-busting. An analysis of 16s rRNA sequences in the 1990s revealed that the microbial diversity of the lake was far greater than expected. The Wisconsin team detected the presence of unidentified bacteria species and possibly archaea around the lake's vents. They measured a range of temperatures from freezing to 248 degrees F. The extremes of temperature provide optimum conditions for cold water microbes, the psychrophiles, as well as hyperthermophiles.

"Analysis of hydrothermal vent water chemistry reveals that not only are the vents in various regions of the lake different, but vents within the same region appear distinct from each other. The chemistry data suggest that each of these vents could represent a different microbial habitat, and thus should have different microbial communities," Maki wrote.

Vent water is the lifeblood of these microbial ecosystems. As water circulates up toward the vents on the lake bottom, it flows over rock heated by that big magma chamber located beneath the lake. Under pressure, the percolating water becomes enriched in carbonate, silicate, and chloride. In some of the areas, Maki reported, the vents are rich in methane, iron, and sulfide. Some of the vents release high amounts of arsenic and mercury into the water. The Wisconsin group concluded that "the interaction of biology with geothermal and geochemical energy may be more ancient than any other ecology."

Absolutely! But I would take it a step farther and say that the interaction between biology and any type of thermal *or* chemical energy is the most ancient of all. What other interaction could exist? Life—

metabolism—depends on chemical energy regardless of its source. *Anywhere* thermal or chemical forces are at work biology could be right there mixing it up.

Without the tools of modern microbial ecology, the Wisconsin group had to make a lot of inferences about the presence of microbial life in Yellowstone Lake. They reported finding "reduced" inorganic compounds of geothermal origin, hydrogen sulfide, iron, and methane in the vent fluids "at concentrations capable of supporting chemolithoautotrophic (geochemical-oxidizing, carbon dioxide-fixing) bacterial growth. Closely linked to the presence of reduced geochemicals was an abundance of chemosynthetic bacteria and dark CO_2 fixation activity."

The 16s rRNA sequences revealed the presence of numerous microbes. But to identify them required matching sequences to those already known in databases. Attempts to culture microbes using standard methods produced only a few species—those darn weeds. The Wisconsin team faced the same problem all microbial ecologists were experiencing in the 1990s and early 2000s with standard culture methods. Algae grew just fine as did organisms called heteroflagellates, which are microscopic eukaryotes that feed on bacteria. Heteroflagellates are a key link in the food chain between microorganisms and everything else from zooplankton to fish.

"This low ability to culture microorganisms extends to just about every habitat that has been studied and has inspired the use of molecular approaches for assessing microbial diversity and ecology," Maki wrote.

Microbial ecologists began using 16s rRNA sequencing to assess microbial diversity and identify microorganisms based on sequence comparisons. The problem was only about 5,000 species of microbes were known to exist. If the sequences did not match any in the database, the microbe was unidentifiable. Labs began using "enrichment cultures" to encourage the unknowns to grow. Thomas Brock defined enrichment culture in his textbook, *Brock, Biology of Microorganisms*, as "Use of selective culture media and incubation conditions to isolate microorganisms directly from nature." Basically this means try growing microbes in mediums based on nutrients in their natural habitats and at the same temperatures and pressures of the microbe's natural environment. This method worked better than off-the-shelf culture mediums like agar, but it still required a great deal of art. That's one reason Stetter also employed optical tweezers to isolate microbes from the samples at his lab. No one isolated and cultured microbes better than Stetter.

Most labs spend many months to successfully isolate and culture a single unknown species from their raw samples. The MATBI, using Diversa's parallel cultivation methods and the DOE's sequencing power, would accomplish the same thing with every species of microbe in a sample in only a couple of weeks.

"The power of applying molecular technology to microbes is astonishing. Consider this: In all of Yellowstone's geothermal features, in over 137 years classic microbiologists named just 40 or so species. In the past 15 years, the molecular biologists [using 16s rRNA sequencing] have named about 500 species, and the rate is rapidly increasing," Varley said.

The Wisconsin team determined that the same sulfur-loving microbes at work in the terrestrial hot springs and geysers of the park appeared to be on the job at the vents on the lake bottom. They found indirect evidence of archaea and bacteria working side by side. The chemical conditions at the freshwater vents were identical in many ways to deep ocean vents, suggesting hyperthermophiles don't care whether they have saltwater or freshwater. Freshwater microbes perform the same metabolic chemistries with sulfur, hydrogen, carbon dioxide, methane, ammonia, nitrogen, and other elements as their deep ocean counterparts. All they seem to need are vent fluids and high temperatures to be happy, just like Stetter!

Vent fluids are rich in minerals and provide plenty of spontaneous chemical reactions to spark the metabolisms of archaea and bacteria to convert inorganic elements into organic compounds for their cells. Even though individual species were hard to identify, the Wisconsin group found abundant microbial life on the lake bottom around vents in the form of yellow and dark banded mats. These microbial mats attract other freshwater life, such as plants, sponges, leeches, and worms that attract cold-loving native cutthroat trout, which dart into the edges of the superheated waters to grab a take-out dinner. A long line of sulfurous-looking mats in the deeper part of Yellowstone Lake are named the Yellow Brick Road. Microbiologist and ecologist David Ward of Montana State University likens microbial mats to tropical rainforests in their ecological complexity.

Nearly every research paper involving Yellowstone Lake's vents and spires published after 1996 has included the name Dave Lovalvo, who designed, built, and operated a series of increasingly sophisticated remotely operated vehicles (ROVs). ROVs are underwater robots equipped with cameras, lights, and various types of sampling equipment and sensors for testing temperatue, salinity, and other factors. ROVs are powered by electric motors with propellers and are connected by a long umbilical cord to a console on a ship or a boat. Lovalvo was one of the first scientists in the world to build these underwater vehicles for marine research. He is considered one of the premier ROV builders and operators, frequently accompanying Robert Ballard on National Geographic Society expeditions to discover and explore ancient shipwrecks. Lovalvo's ROVs have exclusively explored all of the underwater features of Yellowstone Lake. His name appears on all of the studies published in the past five years by the U.S. Geological Survey, which mapped the lake bottom and locations of hydrothermal systems. His work on Yellowstone Lake started in 1985. Significantly, Lovalvo joined the MATBI expedition. Lovalvo shot the video presented by Lisa Morgan at the first Thermal Biology Institute meeting. That video, of course, is what inspired Mathur to plan this expedition.

On August 1, 2004, I flew to Jackson Hole, Wyoming, to meet the Diversa team. The Diversa group stuffed four SUVs full of gear along with 20 scuba air tanks picked up at the local fire station, which doubled as the town's scuba training facility. After buffalo burgers at a local diner, we made a caravan to a lodge near Yellowstone Lake, where the group checked into the park's original rustic cabins.

The next morning we headed straight to the park service headquarters' boat dock for a briefing. Gary Nelson, dive master for the National Park Service's Yellowstone Lake scuba unit, described the diving conditions and sites we would be exploring. Nelson is 100% park ranger, and an expert cold water diver. Diving at a high altitude is tricky by itself. Add hydrothermal vent fields and you have a fairly high-risk situation. I don't think any of us had actually considered the danger. I was too obsessed with the water being cold to worry about vents or altitude. Nelson spelled it out. Number one, high-altitude diving increases the risk of embolism—air bubbles in your blood. Decompression formulas which factor your depth and time spent underwater become more complicated at higher elevations. To minimize risks we dived no deeper than about 50 feet and limited our time on the bottom to 40 minutes. Number two, the body's demand for oxygen increases at higher elevations, so it's easier to get tired and make mistakes. If you aren't in good shape, the exertion that goes into diving can wipe you out. Number three, hydrothermal vent fields are generally unstable, meaning new vents may open suddenly beneath you or nearby vents may unexpectedly belch extra vent fluid along with hydrogen sulfide gas and methane, maybe some extra arsenic too. These burps are called "steam explosions." The vents are positioned randomly along the floor off the West Thumb shoreline, which means one needs to be careful moving through the field so as not to accidentally swim over the top of a vent and boil a hole through your wetsuit and into your chest or groin. The vents are not always obvious. You have to watch for the shimmering appearance of heat being released into the water column, like heat rising off a blacktop road in August.

The sampling team would dive in pairs, equipped with a pole that has a thermometer taped at one end, another pole with cups at the end for scooping sediments, and big syringes for drawing vent water whenever the temperature gradients allowed us close enough to the source. Nelson asked if we had any questions.

"What's the water temperature?" I wondered.

Nelson said it was about 40 degrees F, but maybe five feet from a vent the temperatures can be 60 degrees F. Two to three feet from a vent the temperature can rise to 80 degrees F. Directly over the vents, the temperature can be more than 200 degrees F. That was good enough. If my Gumby suit didn't work, I could cozy up near a hydrothermal vent if my body began turning blue. The fact that the lake sits on top of a magma chamber waiting to blow the northwestern United States to smithereens did not seem to bother anyone.

After the briefing we walked down to the dock where the park service keeps a 28-foot aluminum all-purpose boat called the *Warwood*, which, if painted black, would resemble a craft made for transporting SEALs for a beach assault. It has a nifty plow bow that is made for running straight up onto a beach; and better for us, the bow unlatches to provide a diving platform that allows you to slip straight into water.

Docked in front of Nelson's assault craft was a boat that looked like it was shipped from the set of the Kevin Costner movie *Waterworld*, named *Cutthroat*, after the premier native trout in Yellowstone. Standing on its bow was Lovalvo, the most gainfully employed freelance ROV builder and operator in the world. (See Plate 5.) In addition to the ugly but highly functional *Cutthroat*, he owns several ROVs worth $500,000 to $1.5 million each. He is probably the most skilled ROV operator on Earth. Lovalvo got his start at about age 20 when he bought a one-man submarine for "play." By the early 1970s he was building and driving underwater robots for pleasure and work, and by his own admission, as an obsession. Later, he became a contractor for the Woods Hole Oceanographic Institute as an *Alvin* pilot and the deft handler of *Jason*, the ROV operated for years by Woods Hole for the popular Jason Project. He was also the member of Robert Ballard's team who discovered President John F. Kennedy's PT-109.

Lovalvo had just returned from the *Titanic* with Ballard, the National Oceanic and Atmospheric Administration, and Roy Cullimore, who was back studying the rusticles. How about that?

Bob Dylan: "Blame it on a simple twist of fate."

The expedition's first target was the spires, the series of about 15 10,000- to 12,000-year-old structures of mysterious origin. If the spires were on dry land with boulders laid across their tops you would have something similar to Stonehenge. The spires are only about 200 yards from shore in water no more than 30 to 50 feet deep. By the time of our expedition, only the National Park Service scuba diving team had been down on the spires—officially. Recreational divers are discouraged from diving on them, but no doubt some divers have done so surreptitiously.

Lovalvo and his ROV first explored the spires in 1997 with the University of Wisconsin group. (See Plate 5.) "The visuals were stunning. Through the dim green 'fog' of somewhat turbid nearshore water ghostly shapes emerged; up close, it suddenly became obvious that they were towering columns. Among the lot, graceful individual spires loomed like stalagmites, with clusters of spires resembling ancient castles interspersed among the string," Russell L. Cuhel wrote in an article "The Bridge Bay Spires: Collection and Preparation of a Scientific Specimen and Museum Piece."

"Looming in the camera's lens, the structures varied from mere nubs to towers over 15 ft high, many covered with luxuriant growth. Well infused with natural sunlight at this depth (45–60 ft), large populations of algae covered the sides and tops. As we were to discover, a variety of animals, including colossal examples of freshwater sponges, also make the spire surfaces home," Cuhel wrote.

The Wisconsin team returned in 1999 in a follow-up study funded by the National Science Foundation. National Park Service rangers Wes

Miles, Rick Mossman, and Nelson donned scuba gear to retrieve the first sample of a spire that had broken off and was laying on the bottom. The dive team found a two-and-one-half-foot-tall spire segment and brought it to the surface.

The internal structures were not like chimneys seen at deep ocean vents. The spires obviously channeled vent fluids and gases when they were active, but they differed significantly from smokers, which scientists postulate are created by geological processes that involve the gradual buildup of minerals. Black smokers are typically made of iron and sulfur minerals. White smokers contain a lot of calcium and silicon. All smokers have a central pipe or chimney through which the hot vent fluid streams like a small underwater volcano. The spires are nothing like that.

The Wisconsin team took their spire sample back home where it was given a computed tomography (CT) scan at a hospital before they cut it open. The team's excitement exudes from their report. They made about 150 CT images that revealed an interior with a very-low-density silica mass on one side and a harder "conduit-like structure" on the right with areas that resembled iron-sulfide residues. In the single respect of the conduit it was similar to deep ocean hydrothermal vents. A thin veneer of reddish-brown material surrounded one conduit. I immediately thought of the paste I held in my hands aboard the *Keldysh* that came out of a rusticle. Unlike smokers, the conduit did not extend to the top of the spire. It traveled only halfway up. The upper half of the spire was the most porous, low-density portion of the structure, resembling the vascular organization of rusticles.

"No single mechanism appeared to explain the structure; rather, it appeared as if a combination of geochemical and geophysical forces worked to shape the object," Cuhel wrote in the 2002 report. "The spongy, porous, fragile fragments aroused substantial excitement: these were not at all like the hard pipes we had so often collected with the submersible! Clearly different mechanisms had been involved in the creation of these spires."

Indeed. Chemistries of the spires conducted by the U.S. Geological Survey (USGS) suggested that microbes of the varieties that processed iron and sulfur may have been primary residents, perhaps even builders of the spires. Oxygen isotope studies suggested the spires were formed at temperatures between 175 and 250 degrees F. I noted a comment in the USGS report with equal interest: "At least 8–10 spires up to

7 m tall consist of diatom-rich areas and fibrous masses and globules of amorphous silica that could be of microbial origin."

No one speculated how the spires came to be or what processes were involved in building them. But it was clear that they were not hydrothermal vent chimneys. I had studied the detailed diagrams of the rusticles made by Cullimore and Pellegrino a dozen times. I'm not a scientist, clearly, but like anyone, I'm entitled to a theory or two. I wondered if the spires were giant microbial fossils—remnants of an ancient microbial Atlantis. After seeing the rusticles, anything seemed possible with microorganisms given the right conditions and opportunity.

Mathur, Short, Christoffersen, and I rode with Lovalvo to the site aboard the *Cutthroat.* We searched for the spires first with Lovalvo's side scan sonar and GPS. He'd been there so many times it took less than 10 minutes to locate them. We placed buoys to mark the locations of the spires for the divers. After we anchored, the *Warwood,* which transported all the tanks and dive gear, pulled alongside and tied off next to us. Park biologist Phil Doepke joined the group with a third boat. Lovalvo deployed the ROV to give Jay an idea of what we would be looking at and to show us where Nelson wanted us to take samples. The ROV was fitted with a bucket sampler to scoop sediment samples at the deepest point in the lake. Even though the sky was cloudy, the cold water provided a good 20 feet of visibility. Plankton in the shallow surface layer of the lake gives the water the distinctly green cast described by Cuhel. Within minutes, the first images of a spire appeared on a video monitor. The spires really do look like ancient columns or the tops of some Atlantean castle buried deep in the sediment. (See Plate 5.)

After another quick briefing, Nelson gave us the go-ahead to suit up. I donned my suit and inflated my buoyancy control device (BCD)—the inflatable vest and harness that hold the air tank. I added some extra weights to pockets on the BCD to counterbalance the thick neoprene of the wonderful nine-millimeter Xcel. Jay and Eric buddied up, both wearing seven-millimeter suits over skins. Steve Briggs and his wife, both with drysuits, formed a second team. Nelson, Leif Christoffersen, and I made up the third team. Nelson had a drysuit and Leif was in ancient, frayed, seven-millimeter wetsuit that could not possibly keep him warm. He was one tough nut. After everyone was breathing hard from pulling on their gear, Nelson yelled "pool's open," which meant time to dive.

Mathur and Short slipped into the water first so the ROV could follow them and record video used for presentations at scientific and investor conferences. Those are the "hero shots" that wow either type of audience. Nelson, Christoffersen, and I were tasked with doing the actual work: bringing back a large piece of a spire. I hauled a big tub and mesh bag for samples. The cliché here is to describe the spires as ghostly, or eerie silhouettes backlit by a clouded summer sky. All I can think of to say is they are downright bizarre. Their grouping is quite curious. Why, with a lake full of vents, did these spires arise here at one time in the lake's 12,000-year history? As we approached them I thought back to the last ice age before the lake had formed. The spires have not been dated. They are presumed to be about 10,000 years old because that was when the lake is believed to have thawed. But they could be much older, predating the end of the last ice age. Scientists studying the South Pole have discovered more than 75 lakes beneath the permanent ice shield, including Lake Vostok.

A powerful vent system opening on the bottom of a glacier could have melted ice and created a freshwater pool beneath a thick glacier. I could envision an isolated biome of meltwater, perhaps 100 feet in diameter and up to 60 feet high. At its bottom are hydrothermal vents. Overhead is a dome of ice. Such conditions could have created a sudden opportunity for iron- and sulfur-reducing archaea and bacteria.

I thought again of Martinus Beijerinck studying the behavior of microbes in the late 1800s. He and Dutch biologist Bass Becking both believed, "Everything is everywhere, but the environment selects."

The spires might have arisen in isolation before the lake developed with the warming climate. No one knows how long it took for the spires to reach their 50-foot pinnacle or when they became dormant. The empire of the spires would have collapsed soon after the hydrothermal vent system that nourished them shut down. But something about the spires seemed oddly similar to the monstrous consortium of microbes I witnessed on the *Titanic.* The spires could have arisen quite quickly, thrived for a few hundred years, and then died when the vents stopped feeding them. The rusticles have attained a mass of at least 650 tons with more than 6,000 miles of internal surface area in less than a century. Rusticle Park will continue to thrive until the *Titanic* is reduced to ore. Then it will collapse and the remaining microbes will return to subsistence farming on the abyssal plain. Carpe diem.

Based on the general behavior of microbes we can assume that for the spires opportunity knocked and the microbes rocked. Here's a possible scenario. At first the iron-reducing and sulfur-reducing microbes inhabited small geothermal chimneys at the new vent cracks. As the microbial community grew in size and in the number of species, it incorporated itself into the silicate chimneys. Over time, the consortium *became* the spire—a living, respiring entity. The spires look quite similar to inverted rusticles.

I kept thinking of Cullimore's description of the rusticles: "Included among these structures are internal water channels, water reservoirs, hardened iron-rich plates, surface ducts which pass through the external hardened plates and connect the rusticle interior to the outside environment, porous sponge- or pumice-like layers, bundled fibrillar clumps and elegant thread-like 'girders,' which appear to be strung through the structures and channels and apparently provide a measure of mechanical stability to the entire rusticle."

Just like rusticles, the spires have internal water channels, surface ducts, and an outside environment of porous, sponge-like, silica layers, and internal clumps of "silica that could be of microbial origin." They certainly have stability. Nelson and I drifted over to a spire where he motioned for me to collect a two-foot section. It was surprisingly dense and heavy. I floated just above the bottom, carrying it in my arms like a baby. I felt for the basket and placed the spire safely inside. The water became turbid. So many divers—and the propellers of the ROV—stirred the bottom into a flurry of sediment that reduced visibility to zero. Nelson and I were less than two feet apart but could no longer see each other. We worked by touch alone, digging up more pieces of fallen spires out of the sediment. I lay flat on the bottom and began reaching for fragments that were deeper and perhaps better preserved in an anoxic layer of mud. After about 20 minutes, Nelson and I bumped into each other at the basket. It was full, so we each grabbed an end, slowly surfaced, and kicked back to the boat. Leif was waiting for us at the surface.

The following day, the Diversa team dove on the hydrothermal vent fields to collect fluid samples for the study. The vent fields are in shallow water, less than 15 feet deep. The shallow depth allowed us to stay under as long as the air in our tanks lasted. Lying on the bottom in a cold spot between vents, I took a few minutes to gaze at the dozens of shimmering plumes surrounding me. My mind traveled back 4 billion years to when much of Earth was covered by a shallow ocean. This scene closely characterized much of our planet back then. The earliest chemical trace of life is dated to 3.8 billion years ago. But life quite likely flourished much earlier on Earth. Ancient versions of archaea and bacteria would have been mining vent fluids in shallow oceans and freshwater lakes. They would have been building empires—acres of thick, microbial mats as far as the eye could see. With no known predators microbes could have literally blanketed the Earth.

Based on the cooperative and complex organization of the rusticles, and judging from the architecture of the strange spires, I can imagine microbial consortiums of all shapes and sizes with vast circulatory systems and channels. Members of the consortiums would have communicated through immense and intricate webs of capillaries using hormonal signals, cell to cell, to monitor their environments and mobilize against abrupt changes in gases, temperature, and pH. Early life would have been continually subjected to the opening of new hydrothermal vents, the loss of existing vents, volcanic eruptions, and bombardment by asteroids. The forces of natural selection would have been wildly and madly engaged. We know that nascent life forms exchanged genes. The chaotic environment of early Earth could have fostered a great deal of gene-swapping behavior. Could Woese's pre-cellular life have signaled the need to share genes to promote its survival, or would random combinations of genes, swimming in an amorphous mix of primitive, evolving metabolisms, simply click under the right circumstances and take off on the path to cellular-hood? I wondered about Stetter's ancient *Nanoarchaeum equitans*, and its ability to copy the metabolic genes of other archaea. That strange little microbe must be holding some very old secrets. What a marvelous epoch this must have been!

Lying there on the bottom, I became vaguely aware of the sensation of heat against my thigh. I had drifted over a plume of vent water while daydreaming. I turned and saw the other divers making their way back to boat. I secured my sample bottles in the big pockets of my BCD and returned to the present. Varley wrote a report of the expedition's results for Yellowstone National Park, and the study was regarded as a whopping success. As everyone expected, the biological diversity of the lake was much more extensive than scientists had been led to believe by standard culturing. The new samples from the sediments at the vents and the surrounding cold waters yielded more than 250 new microbial species for the lake. The new discoveries represented all three domains of life. Varley said more species are being identified as the analysis continues.

As the results were coming in, Varley was about as excited as he had been in his career at Yellowstone: "Clearly on par with the first wolves to hit the ground in 1995," he said. (Varley headed the project to reintroduce wolves to Yellowstone and later to sequence their DNA with Diversa.)

For the first time, archaea were documented in the lake: 103 species, including several clusters of closely related species that appear to be new to science. Varley, Short, and Mathur expected to find archaea. But they were all "blown away" by the discovery of 27 previously unknown species of *Nanoarchaea*. Until then, only Stetter's single species of *Nanoarchaeum equitans* was known from the published literature. Only two other possible "sightings," both in marine environments, off the coasts of Indonesia and Iceland had been reported. This suggests *Nanoarchaea* may be common parasites or even symbions in some unknown fashion of archaea.

"Identified separately was the newly discovered group *Nanoarchaea*, which increased the worldwide number of species from a single named species, Karl Stetter's *Nanoarchaeum*, to about 28 unnamed lineages. The new species reported here are the first freshwater records of the group," Varley said. The study also identified 71 new species of bacteria and 28 new species of eukaryotes.

However, a number of scientists would slap me silly if I did not point out that Yellowstone's 27 new *Nanoarchaea* will not be officially accepted by the scientific community until they are written up in a formal paper and published in a peer-reviewed journal. So far the results have been published only in a report. Nevertheless, Varley's enthusiasm has not dampened, and the MATBI succeeded in demonstrating the importance and the power of the methods used to conduct the research.

"The remarkable thing to me is that in two days of sampling and about a month in the lab, the molecular guys came up with about 250 new species. The classical biologists in over a century enumerated about 263 species, and there were only two species in common from each list. It's fabulous. Think of what we could do in a summer of sampling," Varley said. "Qualitatively, the newly identified species were a curious lot, ranging from the common to the unbelievable. In domain Eukarya, for example, researchers expected species typical of the lake's subalpine and nutrient-poor character, and indeed one species assemblage was indicative of that type and similar to species in the Laurentian Great Lakes, Siberia's Lake Baikal, and several high-elevation lakes in the Andes Mountains. But they also found marine organisms, species known only from rivers and streams, and still others that are indicator species for nutrient-rich or polluted waters. Until now, one species found has never been observed outside its Antarctic habitat, and another has never been found outside its unusual home in an oxygen-deficient basin in the Caribbean Sea."

The MATBI further demonstrates a key difference between animals and microorganisms. Plant and animal species are defined in part by their geographic boundaries. Evolutionary biologists have shown how species evolve when animals from the same lineage become geographically isolated, Darwin's finches being a classic case in point. Microbe species may be defined less by geographic boundaries than by their metabolic potentials.

"The exploratory MATBI on Yellowstone Lake can unquestionably be considered a powerful and new biodiversity assessment model, which melds classic Linnaean taxonomy with cutting-edge genomics inventories," Varley said. He believes the approach ought to be applied to species inventories for the microbial diversity of all national parks. Meantime, further analysis of the data of all the new microbes in Yellowstone Lake "should identify previously unknown energy pathways in the lake ecosystem."

Read between the lines on that last comment. Recall that PCR was made possible by a microbe from a Yellowstone hot spring. What if the National Park Service had had a biodiversity agreement back then and had been receiving a penny or two in royalties on every PCR test done in the country? The agency could easily fund the MATBI studies of every park. It could build its own JGI-style sequencing center and its own Diversa-style massive parallel microbial cultivation warp drive. Career scientists in the National Park Service would be easily spotted in a crowd as the people with an unmistakable look of serenity on their faces.

"At the very least, the MATBI model will better assist managers in their efforts to conserve biodiversity, the vast majority of which consists of the small organisms that remain the largest void in the story of life on Earth," Varley said.

Until Yellowstone's spires, I had seen nothing remotely similar to the rusticle consortium. Nor had I read anything in the scientific literature about any cousins of the rusticles among the extremophilia. Geologists still believed the spires were made primarily by geological processes, admitting only that bacteria may have lived in them. Based on traditional thinking, bacterial residues in the spires could simply be the by-products of the buildup of dead microorganisms that lived on or inside the spires. No one would come right out and say microbes built the spires. But after seeing them, holding them in my hands as I had done with the rusticles, the hair on my neck stood up every time I thought about it. Months after the expedition, I decided to ask Varley what did he *really* think about the spires?

"A few people have said they might be some type of giant sponge. The first time I saw them I immediately thought of coral. But no coral expert that I talked to would have anything to do with that idea. I still think they look like coral. Whatever they are I believe it's quite possible that they were made by microorganisms," Varley said.

I told Varley everything I had learned about the rusticles and asked his opinion. He paused for a long moment. I thought he was going to tell me I should stick to reporting. Then he said Cullimore's description of rusticles is similar to odd formations at Mammoth Hot Springs in the park that were "designed, engineered, and built by microbes using the chemical elements in the vent water. They are architectural wonders."

I explained to Varley that in my most recent conversation with Cullimore, he had said the *Titanic*'s rusticles were similar in their architecture to formations found at Lechuguilla Cave in New Mexico and in other caves as well. Speleothem is the general term used for cave formations. Geologists have taught us that stalactites, stalagmites, columns, cave pearls, and moon fingers, to name a few, are the "result of interactions among water, rock, and air within caves. As water seeps through cracks in rock, it dissolves certain compounds; for caves these compounds are usually calcite and aragonite (both calcium carbonate), or gypsum (calcium sulfate). Over tens of thousands of years, these drops cause speleothems to form."

I like geologists. They remind me of park rangers except perhaps a little wilder at parties. But they've had us looking at rocks and minerals and seeing nothing but rocks and minerals. It turns out that's not quite right. Microbes have actively shaped the planet we live on and continue to do so when we don't get in their way. A number of natural geological processes are actually the handiwork of microbes, especially consortiums of them. Several types of cave formations turn out to be living consortiums of microbes and minerals. Scientists call these things "biothems." Imagine that. Living rock.

Diane Northrup of the University of New Mexico, Susan Barns of the Department of Environmental Molecular Biology at Los Alamos National Laboratory, and their colleagues reported in 2000 that microbes are involved in the growth of carbonate and silicate cave formations. The microbes have metabolisms associated with sulfur compounds, iron and manganese oxides, and saltpeter. The microbes are able to precipitate minerals either "passively by acting as nucleation sites, or actively through the production of enzymes or substances that lead to precipitation by changing the microenvironment." Microorganisms also can dissolve carbonate and silicate to create cave features via their acidic metabolic by-products.

"While geo-microbiological interaction studies in the outside world are rather common, such studies in caves are just beginning. Studies of geo-microbiological interactions in caves can shed light on basic mechanisms of dissolution and precipitation by microorganisms, and, thus on the origin of specific types of speleothems," the group reported.

Guadalupe Mountain caves are a focus of studies of living cave formations. At the Lechuguilla and Cottonwood Caves scientists found biothems with features known as webulites and u-loops, which appear to be calcified filamentous microorganisms. Slimy formations, called "snotties," were found inside Cueva de Villa Luz, a cave with active hydrogen sulfide vents. The bacterial snotties, which hang from the ceiling, are mineral formations covered with thick mucous-like biofilms of bacteria and sulfuric acid with a pH of 0.3. These odd microbial colonies could become lithified later in the evolution of the cave, producing the u-loops formation seen in the Lechuguilla Cave today, according to Northrup and Barns.

Geologists have assumed the mucous residues are the product of abiological, inorganic processes.

"The physical resemblance of u-loops to living structures in an active hydrogen sulfide-dominated cave, Cueva de Villa Luz, Tabasco,

Mexico, has fueled further speculation that these may represent preserved and living examples of the same microbial/mineral/ mucopolysaccharide structures," Northrup said.

The active Villa Luz system produces prodigious quantities of hydrogen sulfide and other gases, including relatively high levels of carbon dioxide and low levels of oxygen. It is an extremely acidic environment that is rapidly eating away the rock and precipitating minerals such as sulfur, gypsum, selenite, and pyrite. Based on 16s rRNA sequencing, the microbes in the acidic mucous appear to be related to *Thiobacillus.* Mineral formation is occurring in concert with the microbes in microbial mats and stringy structures lining the cave springs, under rocks, and on other surfaces, Northrup said. Sulfur isotope studies indicated the presence of active microbial metabolism of the various sulfur compounds within the cave.

Pool fingers are described as "elongate, pendant speleothems that form underwater. They have an irregular, knobby external appearance underlaid by laminations of dark, clotted micrite and/or clear, inclusion-rich calcite spar. In some samples, the inclusions are obviously bacterial. The external and internal fabrics of Hidden Cave pool fingers strongly resemble well-described microbialites from surface environments. This similarity, combined with scanning electron microscope identification of fossil bacteria in etched samples, suggests bacterial involvement in the formation of the pool fingers," Northrup and her colleagues reported.

Varley is a careful, respected, and experienced scientist. He said it is possible that the spires, the formations at Mammoth Hot Springs, and the biothems are related. After another long pause he said, "I believe there is much more going on with microorganisms than we have imagined."

A growing number of molecular biologists are taking a keen interest in the relationship between biology and geological and geochemical processes. Lesley Warren at the School of Geography and Geology, McMaster University, Hamilton, Ontario, and Mary E. Kauffman at the Idaho National Engineering and Environmental Laboratory published a paper in *Science* titled "Microbial Geoengineers." They began by stating:

> Recent studies of deep-sea hydrothermal vents, highly contaminated, abandoned mines and Earth's deep subsurface underscore the ubiquitous

> presence of microbes in the geo-sphere. A session at the Fall meeting of the American Geophysical Union highlighted the close linkages between microbes and geochemistry. Microbial activity is increasingly implicated in aqueous geochemical processes such as mineral precipitation and dissolution, contaminant degradation, sequestration or mobilization, fossilization, and weathering. These linkages are found in a broad spectrum of aqueous systems, including marine, freshwater, groundwater, and subglacial melt. However, we are only just beginning to understand how microbes and geochemical processes interact. The geochemical reality of these interactions, which often occur at the micrometer scale, has only recently become quantifiable with high-resolution methods such as x-ray absorption spectroscopy.
>
> Unlike higher plants and animals, microbial evolution is one of metabolic diversity, rather than cell complexity and organism structure. Microbes are not restricted by geographic barriers, but their metabolic pathways are necessarily constrained by the available redox couples in the biosphere. However, microbes can use a wide range of electron acceptors other than molecular oxygen for respiration (such as carbonate, ferric iron, nitrate, and sulfate). The geochemical influence of microbes therefore extends to all the major elemental cycles, particularly those relevant to life on Earth. Further, microbes are reaction accelerators, catalyzing otherwise slow redox reactions to kinetic rates that make them of geochemical interest. Thus, microbial influence on geochemical processes is not only widespread, but likely predictable, given a more systematic understanding of the mechanisms and controls involved.

Microbes are geo-engineers. They have been for a long, long time. How much of what we assumed to be natural chemical and geological processes are actually caused by or at least shaped by microbes? Microbes also are bioengineers. To what extent have microbes shaped so-called higher organisms such as plants and animals? What else are microbes capable of engineering?

Pellegrino described the rusticles as more complex than a sponge, which has its own phylum and is considered a primitive animal. In 2002, Cullimore, Pellegrino, and Johnston formally proposed that rusticles, and possibly other consortiums, be classified as independent forms of life. So far, no scientific body has taken them up on the proposal. But attitudes are shifting. Microbiologists already joke that microbes fostered the evolution of plants, animals, and humans to enslave us as protective and nurturing hosts. We are covered and filled with microorganisms. Quite literally, our bodies are complex consortiums—"woven socks of slime" made up of eukaryotic cells that cooperate with multiple species of bacteria and archaea.

"Genomic and evolutionary analyses show us that we are not the single 'individuals' that we think we are. Instead, we and other complex organisms are composed of an interconnected ecosystem of eukaryotic and prokaryotic cells whose interactions can best be understood in the context of community ecology," wrote Edward Ruby and Margaret McFall-Ngai of the University of Hawaii, and Brian Henderson of London's University College.

An estimated 15% of the human genome contains viral and bacterial genes. Does this mean we are part human, part microbe, part virus? Are the rusticles, biothems, and spires something else that microbes can *become*?

Michael S. Gilmore and Joseph J. Ferretti of the University of Oklahoma view the consortium of microorganisms living in the human intestinal tract as a structure worthy of being called an organ right along with our hearts and kidneys.

"Like any other organ of the human body, the gut flora of higher animals—including humans—is a complex association of cells that collectively performs essential functions. Not only does the highly evolved gut flora community extend the processing of undigested food to the benefit of the host, but it also contributes to host defense by limiting colonization of the gastrointestinal (GI) tract by pathogens. The GI tract consortium is, however, unique among organs in important ways. It contains more cells than the rest of the human body. The cells of the gut flora consortium are diverse, consisting of at least 500 different microbial species (most of which have yet to be cultured in vitro), and its composition varies with the organism's age, diet, and health status," Gilmore and Ferretti wrote in a perspective in *Science*.

I already mentioned how the rusticles made my stomach flip a little. I couldn't get over the point made by Cullimore and Pellegrino that the rusticles appear to be something emerging, becoming more complex. It is not us, but its circulatory system and metal slime tissues are too familiar. Was there some Carpe Diem gene present in life from the beginning that drives it toward greater complexity given the opportunity? How odd it is that microorganisms organize themselves into something that appears to be a precursor to arteries and organs—even with iron and rock.

The Bridge Bay Spires, Rusticle Park, and many of the aesthetically pleasing cave formations may be biological entities, distinct from colonies and consortiums. A number of scientists have no doubt about the rusticles and cave formations. Varley and others know the formations at Mammoth Hot Springs and the spires are something quite strange. But the paths of these scientists have not yet crossed. My fantasy scientific conference is to get all of these people together to look at each other's data, and then to converge on the spires. I think it's safe to say that the *thing* indeed has cousins. Now we can ask what do they have in common?

Rusticle Park started as a boomtown, a successful colony established on a desolate plain where life had done nothing more than scratch out a subsistence living over eons of time. With the *Titanic* shooting like a massive asteroid to the ocean floor, the environment was drastically and suddenly altered. Microbes and nutrients buried in sediment were literally tilled from the abyss and found themselves in the midst of a substantial piece of steel and other metals that fairly quickly began generating an electrical current. The electrical current got the ball rolling for oxidizing the iron and provided power for multiple microbial metabolisms. It was fat city. Species that possessed a knack for mining sulfur became rich along with the iron workers. The *Titanic* offered so much work for different species that all of the subsistence farmers of abyssal plain could get a job and make quite a successful living. Even the eukaryotic slime, not ordinarily found in the abyss, was incorporated into the new society. Over a relatively short period of time, the consortium began to merge into something more complex, greater than the sum of its parts.

The spires may have also started as a boomtown. The biothems are booming at their own geological pace. Microbial mats form rapidly wherever new hydrothermal vents appear. What all of these *things* appear to share is opportunistic and cooperative behavior. Studies of consortia date back decades. But only since about 2003 have the tools offered by molecular biology permitted scientists to begin exploring cooperative microbial behavior *in situ*—in the wild. So far, microbial mats have attracted the most attention.

Craig L. Moyer of Western Washington University has studied microbial mats extensively, especially on deep ocean hydrothermal vent systems. Scientists have little doubt that microbial mats played an important survival role for anaerobic microorganisms during the Oxygen Catastrophe, the period in Earth's history when the atmosphere changed from primarily carbon dioxide to oxygen, wreaking havoc on the anaerobic microbial empire that prevailed until about 1.4 billion years ago. Modern microbial mats often consist of consortiums that form a layer cake of anaerobic microorganisms at the bottom and oxygen lovers at the top. In the middle are species that can either use oxygen or convert to anaerobic respiration. These layered cooperatives were in place before the atmosphere began to change 2.5 billion years ago. Before the Oxygen Catastrophe, scientists presume that anaerobes dominated the upper layer. Fossil evidence shows microbial mats have been around for at least 3.5 billion years. Rock fans will recognize that stromatolites are fossilized microbial mats.

"By far the most extensive microbial mats discovered [at hydrothermal vents in the Pacific Ocean] have been those found at NW Eifuku Volcano. These microbial mats (mainly composed of bacteria) are easily distinguished into two general categories. The first and most often seen are the yellow and orange 'fluffy' microbial mats. These fluffy mats occur at many of the diffuse-flow venting sites, which generally have only moderately warmer temperatures than the surrounding sea water. These mats can also be recognized by their ability to form 'bacteria balls' that can be seen slowly rolling downslope after their production in heavily laden mat fields like Yellow Top. They can also be extruded from nearby cracks and fissures from especially steep areas of diffuse venting," according to Moyer's description.

Similar to the snotties, these bacterial balls consist of a sulfuric acid mucous. The mucous helps the vent microbes capture nutrients from the vent fluids, which are primarily reduced iron and sulfur compounds. The mucous also keeps oxygen levels low so that anaerobes can optimize their chemolithoautotrophic metabolisms. Aboard the submersible *Alvin*, Moyer saw that microbial mats also may form. I found the following observation by Moyer especially interesting: "The mounds are covered with a thin crust and small chimneys that are byproducts of this oxidation process, which is similar to the formation of rust. The thin crust acts much like a thermal blanket, retaining the heat along with the vent fluid nutrients. Similar types of microbial mats have been identified at Loihi Volcano in Hawaii, where they were dominated by iron-oxidizing bacteria. Consummate imperialists, species of bacteria have adapted to survive in virtually every habitat on the planet, from the deepest abysses of the ocean floor to the thin air of the stratosphere."

The versatility of microorganisms is associated with the ability to exchange genes with other species. This may be one key factor that permits cooperative, even synergistic behavior—"You scratch my membrane, and I'll scratch yours." Moyer says this gene-swapping behavior in consortiums makes the taxonomy of microbial communities especially difficult.

"The challenge to traditional taxonomy is further compounded by the proclivity of bacteria to commingle and functionally interact in symbiotic living situations. If these traits resist structural and genealogical forms of classification, they also suggest another approach, that of description along the axes of functional interaction. For it is these dynamic inter-relationships that structure bacteriological communities themselves," Moyer said.

He seems to be suggesting, like Cullimore and Pellegrino, that complex microbial consortiums may require their own taxonomic classification. Living rust with its own arteries and immune system, ancient spires in Yellowstone Lake, mucous-covered biological cave formations, and bacteria balls on the ocean floor certainly suggest a complexity that has not been appreciated until quite recently.

Lynn Margulis described microbial mats in shallow regions of the ocean as "sticky, textile-like expanses of mud, usually found in intertidal regions where the strong wave motion of the open ocean is minimized by a barrier of dunes or rocks. Mats are often spectacularly colored purple and green as a result of massive populations of photosynthetic bacteria. . . . Mats usually flourish in areas that are too salty or stagnant for larger organisms."

The uppermost layers of the mat contain blue-green bacteria, which are capable of photosynthesis; that is, they use the energy of light to split hydrogen molecules from water and hook them onto carbon dioxide molecules. In this way they manufacture organic compounds needed for their bodies, as well as produce oxygen. Immediately below this food-producing layer are several distinct strata of oxygen-tolerant bacteria, which consume the food and oxygen produced above them. Located below and protected by this layer from the blue-greens are bacteria, which require light but are poisoned by oxygen. According to Moyer, microbial mat communities, through symbiotic production of food and micro habitat, survive and grow for hundreds of years. Large expanses of living mats similar to those described by Margulis still flourish in shallow ocean regions of the Caribbean Islands, Australia, the southeastern United States, and Baja, California.

"Syntrophic relationships" is a term increasingly used in the scientific literature to describe the behavior of microbes that form consortiums. The term is derived from the word "syntropy," which is credited to Buckminster Fuller, known for the geodesic dome, his open-mindedness, and 28 books. According to a biography of Fuller, he attended Harvard, "but was expelled from the university twice: first, for entertaining an entire dance troupe; and second, for his 'irresponsibility and lack of interest.'" Fuller, who invented the word "synergistic," is credited with popularizing the term syntropy. Wikipedia provides this definition: "A tendency towards order and symmetrical combinations, designs of ever more advantageous and orderly patterns. Evolutionary cooperation. Anti-entropy."

Blessed are the microbes for their syntropy. As we gain a better understanding of them, we will need to start thinking bigger—from microbial colony to microbial cooperatives and right on to alien-like *things.* The behavior of these complex microbial entities is strikingly familiar. They organize themselves in similar patterns and coordinate their metabolisms with incredible efficiency. They seize every opportunity to become something greater, something more. We know they exist. They have names like rusticle, spire, and biothem. The question is, what are they?

7
CARPE DIEM

Archaea inspire theories on the origin(s) of life and the search for extraterrestrials.

Picture yourself in a boat on a river with tangerine trees and marmalade skies.
John Lennon

The long blades of a Russian MI-8 cargo helicopter beat phat-phat-phat-phat through the thin air above the Vostochniy mountain range. Lush, forested mountaintops streak less than 500 feet below the belly of the chopper, which ferried a team of American and Russian scientists from Petropavlovsk 160 miles north to the Valley of the Geysers. Discovered in 1941, the valley is a primordial crack in the planet's surface, venting clouds of hot sulfur gas and spewing streams of boiling acid.

The men and women of the team, bundled in colorful down-filled jackets, sat with child-like anticipation on gear bags and slender aluminum benches lining the sides of the cargo bay. The air inside the helicopter was cold. The whine of the twin turbine jet engines was deafening. The strong odor of fuel and mildewed canvass burned the nostrils. No one cared. It was already late afternoon, and everyone was simply grateful to be en route to the gates of hell.

Chaos and communication breakdowns, typical of expedition travel, kept us waiting at the heliport on the outskirts of Petropavlovsk, a sad decaying city that once teemed with commerce as the former

Soviet Union's secret nuclear submarine base of operations. But delays began before the U.S. team's Magadan Air flight from Anchorage could land at Petropavlovsk's Yelizovo airport. Fittingly, the city was cloaked in fog as we approached just after dawn. Instead of landing, the aging Tupolov passenger jet circled mysteriously over the shrouded city for two hours. Finally, a flight attendant admitted the instruments required to navigate a landing through thick fog were broken. She smiled. With a shrug of the shoulders and wave of the hand, the Russians keep on truckin' despite the decline of their empire. And, they did not take credit cards for helicopter flights, as the American expedition organizers discovered after arriving at the heliport. We had to cough up $7,000 cash, adding another delay as we raced back to Yelizovo and nearby banks to empty ATM machines. All of this for the privilege of crowding into the cargo bay of a decommissioned military MI-8.

Nevermind the Russian workhorse's hair-raising civil aviation safety record: an average of seven crashes a year, killing 180 passengers between 1998 and 2004. *Pravda* reports accidents are usually caused by "piloting mistake, improper technical servicing, or violation of operating rules." Twenty minutes into the final leg of our journey a couple of team members looked pale and were too quiet.

Vera Dmitrieva, our team's rouge-cheeked, motherly host, and a former Soviet scientific translator and biotechnology business expert, shouted above the turbojets, pointing excitedly to the Karymsky volcano through a small view port. The Karymsky is among the largest of more than 100 volcanoes rising from the Kamchatka Peninsula, one of the most geologically active regions on Earth; the Pacific Ring of Fire. The peninsula is the size of California and populated by only 300,000 people. All but the indigenous population lives in Petropavlovsk.

Kamchatka is wild: spiked volcanoes and deep glaciers evoking visions of a primal planet. Much of Earth might have looked this way as early as 4.3 billion years ago. Our MI-8 was beating an aerial path directly to an immense concentration of fumaroles, hot springs, and geysers, where colonies of archaea were discovered in August 2000. The Kamchatka microbes are related to species found in the hot springs of Yellowstone National Park in the United States, the Wai-O-Tapu (Sacred Waters) geysers of New Zealand, the volcanoes of Iceland, and inside deep ocean hydrothermal vents around the globe.

Aboard the helicopter is an interesting scientific team with scientists from Diversa, the Department of Energy (DOE), academia, and

VECTOR, the former Soviet Union's main bioweapons laboratory. We came in August 2004 in order to gather microbial samples from this amazing place. The usual suspects were there: Short, Mathur, Christoffersen, and Stetter, and I, too, was a regular by then. James Noble, the bearded soft-spoken director of the Initiatives for Proliferation Prevention (IPP) program at the DOE, was there. His genuine passion is helping former Soviet bioweapons researchers find honest work since the fall of the USSR. Noble has the tolerant personality one must develop over years of navigating federal bureaucracies. He seems thoughtful and a bit shy. He had recently undergone a cardiac procedure but was not going to let that keep him from getting out of Washington on an adventure bearing the fruits of his program. The IPP is the federal partner in the United States Industry Coalition, which is a non-profit consortium of U.S. companies that work with the former Soviet Union on non-proliferation of weapons of mass destruction (WMD). The IPP has been helping Russian microbiologists since the early 1990s to launch a commercial biotechnology sector. Not only is it critical to the new Russian economy and scientific community, but it is in every Western nation's interest to help former bioweapons scientists find peaceful scientific projects that pay a decent salary.

Tamas Torok, a slender, quiet, classic academic type from the Center for Environmental Biotechnology at Lawrence Berkeley National Laboratory, which is part of the DOE, was there. Torok has done some of the federal government's key work on rapid identification of anthrax spores. But he specializes in basic microbial diversity and microbial ecology. He's a wonderful lecturer and becomes passionate when talking about the origins of life and the possibility of life beyond Earth. He also has that adventurous streak that takes him to the outer limits in search of unknown microbes. Torok began collaborating on non-WMD microbial projects with Russian scientists during the early 1990s.

Torok is good friends with the team's Russian host Vladimir Repin, the fiery director of the Research Institute, State Research Center of Virology and Biotechnology VECTOR Collection, Novosibirsk, Russia. Repin has a classic shock of thick white hair and an unintentional scowl lined into his rugged face. VECTOR sounds very James Bondish because it is, except it's real. The 100-laboratory facility is located about 20 miles southeast of Novosibirsk. According to globalsecurity.org, VECTOR was producing long-range strategic and operational missile warheads packed with the Marburg virus right up to the collapse of the Soviet Union in 1989. It was a little daunting to look at this man at the front of the helicopter bay and consider that not long ago he was cultivating Marburg virus, smallpox, bubonic plague, and anthrax for missile warheads. In the late 1990s, Iran reportedly attempted to recruit VECTOR's scientists. Repin would have been at the top of their list. In 1993, VECTOR had about 3,500 scientists. By 1998, the faculty fell to 1,500. Who knows how many were recruited. Some have disappeared.

Few scientists begin their careers dreaming of destroying millions of people. While Repin did what he was told by his former government, he was never any less interested in the magic of microbiology than Stetter, Short, Mathur, or Torok. Repin was eager to collaborate on non-weapons science projects. By the time Iran was allegedly headhunting VECTOR scientists in Russia, Torok and Repin were exploring Lake Baikal, the world's oldest and deepest lake, located in Siberia. It has rare freshwater hydrothermal vents similar to those on the floor of Yellowstone Lake. Torok and Repin collected microbial samples for potential medical and industrial biotechnology uses. But they need Western biotechnology companies as collaborators. Diversa became involved with the IPP and Repin in 2000.

Dmitrieva, the executive director of the Center for Ecological Research and BioResources Development in Puschino, Russia, was there. This top-tier negotiator for former bioweapons scientists with western biotechnology and pharmaceutical companies and grandmother of three was delighted to have the opportunity to begin teaching courses on how to blend non-proliferation-oriented science with business for the new Russian economy.

Anna-Louise Reysenbach, a young-spirited, silver-haired microbiologist from Portland State University in Oregon, wearing Teva sandals, bumped into the group at the heliport. She was planning to travel to the Uzon Caldera, the other major sampling site about six miles south of the Valley of the Geysers, where academics from Russia and at least five American universities supplied with National Science Foundation (NSF) and NASA grants have been collecting archaea since 2003. The Uzon Caldera is where Stetter went on his first Kamchatka trip in 2000. The Uzon Caldera and the Valley of the Geysers have only been open to Westerners since the mid-1990s, initially to salmon fishermen and hunters. Fewer than 200 foreigners are allowed to visit this region in any given year.

Reysenbach specializes in the *Aquificales*, which is a large and poorly understood heat-loving group of bacteria. Like many species of archaea, these bacteria primarily use inorganic compounds for their source of carbon and energy. *Aquificales* attach to the surfaces of rocks and grow into beautiful long delicate strands, like locks of yellow and pink flowing hair. NASA was funding Reysenbach to study the metabolism of *Aquificales* as a model for the evolution of metabolism on early Earth. When Reysenbach saw that Stetter, Repin, and Dmitrieva were with the team, she pleaded to join their expedition. Her addition was perfect. The team then had scientists to explore both the archaea and the bacteria of one of the planet's most extreme environments.

As we approached the Valley of the Geysers the excitement level in the group jumped considerably. The helicopter circled a couple of times over the valley, giving us a clear view of where we would spend the next week. The only man-made structures below were a few lodges built of birch, a square wooden landing pad, and an artery of wooden walkways that extended out to the most significant thermal features. The vegetation was Eden-like. A few narrow pathways threaded through the vegetation down to streams from the ridges and mountainsides. After a journey that began nearly 20 hours earlier, we were finally about to land.

The turbine jets whined down to a whimper. The spinning blades slowed to a stop. We landed. The Valley of the Geysers was a startling landscape, a deep lateral incision through the fascia of the planet's thin skin. Clouds of hydrogen sulfide rose from hundreds of vertical cuts on the sides of the steep canyons and spread into a blanket of thick, sultry fog below and beyond the landing pad. The expedition team waited with barely contained anticipation, murmuring with excitement as the pilots came around to open the bay door at the rear of the helicopter. The hatch swung down with a thump, forming a ramp. Repin was up quickly and exited first, retrieving his olive green, weathered pack from the pile of gear inside the bay. The others followed single file, greeted by the smack of sulfurous humidity, a hissing below the shroud of fog, and the echo of rushing whitewater ripping through the bottom of the canyon.

The long winter's snows—up to 60 feet high—had melted and given way to vegetation of a variety entirely unfamiliar. None of the thick, stalked, leafy *Filipendula camtschatica* surrounding the landing pad are less than five to six feet in height. Yellow flowering *Senecio canabifolia* extended across a wide meadow to the south. *Aconitum* with its purple blossoms blended with the lush greenery that covered every inch of land one could see except for a boarded walkway that led down into the fog and a single narrow footpath that beckoned upward into the dense wilderness. Beyond the snowcapped ridges of the steep mountains that encircle the valley was a landscape dotted with more than a dozen smoldering cones rising toward heaven.

Christoffersen and I unloaded the team's packs and the heavy cases filled with portable laboratory equipment and sampling tools. The others stretched tired muscles and achy backs as they waited for their gear on the square wooden platform. Repin straightaway assumed the manner of one clearly in charge. Hierarchies fell quickly into place. Repin cautioned, ordered even, the group to *never* wander up the footpath alone or step off the boarded walkways except when sampling. He then turned and proceeded to several wooden cabins where we would be protected at night from 1,600-pound predators that are the larger version of Yellowstone's grizzly bears. The narrow path that beckoned up into the wilderness was made by bears. We also discovered immediately after filing out of the helicopter that the clean crisp air was swarming with millions of bloodthirsty mosquitoes the size of bees.

Repin occupied a single, small cabin. The Russian family who are the caretakers of the facility dwell on the ground floor of the larger log-hewn cottage. Our guest rooms were located on the second floor. Other than our group, weather permitting, tourists arrive once, sometimes twice daily by helicopter from Petropavlovsk. The tourists are fed fresh salmon and caviar, shown the hot springs and geysers from the safety of the boardwalk, allowed to make their videos, and whisked back out within two hours.

When Christoffersen and I finished unloading the helicopter, most of the group headed to the lodge to find their bunks and unwind from the journey. Short and Stetter left their gear on the landing pad clear of the helicopter and followed the boardwalk down to see what lay beneath the dense fog. Stetter produced a weathered leather hat with an Indiana Jones flavor. He has worn the same hat on sampling expeditions since the 1980s. He also carried his trademark fiberglass pole with the dented tin sampling cup taped at one end. It doubled as a sturdy walking stick and divining rod as the two men descended into the hydrogen sulfide mist. (See Plate 6.)

"Can you believe this?" Short said softly. "It's like walking through the looking glass back to the beginning."

Hydrogen sulfide gas was escaping from fumaroles. Geysers sprayed fountains of boiling acid at intervals predictable only to those familiar with the geyser's schedules. The concentration of hydrothermal systems produces a constant low-frequency rumble below the ground. Hundreds of fumaroles vent vapors from both sides of the steep rocky slopes. Hot springs collect in pools as small as six inches to 12 feet across as far as we could see up and down the canyon. Many have small waterfalls spilling acid down sheer rock faces that glisten with mucous—biofilms, made of multiple species of microorganisms. The low pH of spring water leaches minerals from the rocks, all stained in broad brushes of yellow, orange, red, purple, and black. Living mats, much thicker than the films, produce multiple shades of greens and yellows. The metabolisms of the species that form mats link syntropically to permit greater energy efficiencies and access to resources than a single species can get alone.

In the shallow streams that flow from hot springs, at a safe distance where the water has cooled to around 150 degrees F, and the pH is nearly neutral, long streamers of pink and yellow *Aquificales* wave like grasses in a liquid prairie. *Aquificales* represent one of the oldest 100 or so evolutionary branches of the bacterial domain.

Short and Stetter reached the bottom of the canyon and, of course, at the first opportunity departed from the wooden walkway. We arrived at a broad expanse of wet multicolored rock surrounded by fallen trees, the rotted wood bleached nearly white. The vegetation appeared poisoned. Stetter looked satisfied. We had entered the outer limits. (See Plate 6.)

With unmistakable deviance, Stetter announced, "Here." He waved his staff at the rocks and pointed at the dead trees. "We can wait a little bit and see what happens."

Repin had tipped him off, but he wasn't sharing the secret. Then, the sound of hissing, gurgling, and whoosh! Velikan Geyser, the largest in the valley, erupted about 20 feet away, firing a jet of boiling acidic water nearly 60 feet into the air. Hydrogen sulfide gas began condensing into a cloud above the fountain, adding mass to the gray shroud over the canyon. The startling eruption and the unanticipated enormity of it sent Short and me wisely scrambling backward, but Stetter stood his ground.

"This is *the* life!" he pronounced.

Stetter is microbiology's Lieutenant Colonel Kilgore, the character played by Robert Duvall in *Apocalypse Now*: "Do you smell that? It's napalm, son. Nothing else in the world smells like that. I love the smell of napalm in the morning. . . . Smells like . . . Victory."

Hydrogen sulfide would have been the scent of life declaring victory on primordial Earth.

Gazing up at the cloud formed by Velikan Geyser, I grasped for the first time the impact that the degassing of vapor from the hot young Earth could have had on the early atmosphere—and how rapidly the early atmosphere could have developed given the right conditions. Magma was loaded with water from the beginning. Carbon dioxide and hydrogen sulfide vapors would have been roiling off the molten surface as it cooled. It struck me that clouds, which naturally harbor massive communities of microbes, could have provided the first safe haven for the development of life on Earth. Soupy clouds with all the necessary compounds would provide friendlier temperatures than a hot degassing surface. Why would life have to wait for ponds or oceans and vents?

We wandered along after Velikan settled down, Stetter chose our path along the rocky edge of the rushing stream. No fish live in this part of the river because of the low pH caused by the runoff of the hot springs and vents under the bed of the stream. Stetter stopped suddenly and poked the ground in front of him. The pole punctured a crust. A puff of steam escaped and was swept away by the wind. Acid boiled beneath the hole, teeming with invisible archaea. Stetter looked back and grinned, then chose another direction. Wandering along a path with eggshell-thin crusts and unpredictable geysers took getting used to, but we trusted *Ventwalker*. (See Plate 7.)

Stetter was scanning the narrow valley for the hottest springs to begin sample collecting the next morning, keying on the colors of the surrounding rocks and the distance from the springs of streamers and rock slime.

"This is real volcano country," Stetter said breathlessly. "A paradise of early life. We will find many new organisms, I believe, and also some new *Nanoarchaea*."

Stetter said he felt from the moment he entered the world of archaea that he was getting a front row seat to the drama of very early life. "Based on their volcano-adapted primitive lifestyle I raised the hypothesis that these organisms may be remains of early life on Earth and that similar organisms could have existed already nearly 4 billion years ago. At those times, due to very active volcanism and perhaps a still brittle crust, Earth had been much hotter than today," Stetter said. "Today, the hottest environments are found mainly in areas of active volcanoes along tectonic fracture zones and hot spots. I look for archaea in the heated sediments like these beneath the stream, and of course in the hot springs. I found archaea living in deep sea hot vents with their spectacular black smokers, and, in addition, I found deep subterranean non-volcanic geothermally heated areas that were biotopes with temperatures of about 100 degrees C some 3,500 meters below the surface of the Alaskan North Slope permafrost soil and below the bottom of the North Sea. Within there, Jurassic oil-bearing sandstone and limestone also harbor communities of high temperature organisms."

The metabolisms of first life must have evolved to fit the environmental conditions that existed on the planet. These would have included both hot and cold temperatures. Woese's theory of amorphous cells mixing with pools of genes would allow primitive organisms to assemble tools to tap whatever energies were available.

Carbon dioxide and hydrogen would have been among the most abundant compounds on early Earth. The first metabolisms here might have taken advantage of CO_2 with hydrogen as the electron donor. But it seems likely that several varieties could develop simultaneously from Woese's batch of nascent cells and genetic material. Clues to this must be present all around us in the ecosystems of the valley, in the genes and proteins of all microbes, and in ancient isotopes trapped in fossil sediments around the world.

"Most species of archaea exhibit a chemolithoautotrophic mode of nutrition—inorganic redox reactions serve as energy sources and carbon dioxide is the only carbon source required to build up their organic cell material. Therefore, these organisms are fixing carbon dioxide by chemosynthesis and are designated chemolithoautotrophs. The energy-yielding reactions in chemolithoautotrophic hyperthermophiles are anaerobic and aerobic types of respiration. Molecular hydrogen serves as an important electron donor. But they can obtain electrons also from sulphide, sulfur, and ferrous iron. In some hyperthermophiles even oxygen may serve as an electron acceptor, so some of them were using oxygen at very low concentrations long before photosynthesis," Stetter said.

Short and Stetter returned to the boardwalk and began climbing the steps out of the mist. As they headed for the lodge, I stopped at an overlook to linger behind the looking glass a little longer. Sitting amid the hissing fumaroles, gurgling hot springs, and a pit of bubbling reddish mud below the overlook, I thought back to Yellowstone, diving in the shallow waters of the vent field. That image is forever burned in my brain: clear greenish water, 15 feet deep; shimmering plumes of vent fluids rising in every direction; and microbial mats, mining the fluids as well as each others' metabolites. Parts of the early ocean may have looked much the same, but how long ago? The surrounding valley would have resembled an early surface hot zone, but when?

The most specific age for Earth and the universe was published in 1642 by John Lightfoot in his book *A Few and New Observations Upon the Book of Genesis: The Most of Them Certain; the Rest, Probable; All, Harmless, Strange and Rarely Heard of Before.* What a title! Lightfoot was vice-chancellor at the University of Cambridge. It is unclear how Lightfoot arrived at his conclusions, but he stated in his book: "Heaven and earth, center and circumference, were created together, in the same instant. . . . This work took place and man was created . . . on the 17th of September 3928 B.C. at 9 o'clock in the morning."

Do you suppose this is why many people go to work at 9 a.m.?

I'm not making fun of Lightfoot. We are products of our times, constrained by the limitations of our technology to explore the world around us. Only nine years before Lightfoot's publication, the Vatican released Galileo from house arrest for insisting Earth was not the center of the universe. Leeuwenhoek was 10 years old and would not discover a magnifying glass at a textile shop for four more years.

The currently popular theory of early Earth, known as the Hadean Eon, is based similarly on the limitations of our era's technology to determine when the planet formed a crust and core, an atmosphere, oceans, continents, and life. The oldest rock representing ocean sediment and the oldest chemical evidence suggestive of life are both 3.8 billion years old. Most scientists believe that life dates back to 3.9 billion to 4 billion years ago. That's where most draw the line because of assumptions about the conditions of the planet's first 600 million years of development. But what actually happened between the birth of our solar system 4.6 billion years ago and the first traces of an ocean and life 3.8 billion years ago is a mystery. Much of what we've been told about this period is scientific mythology.

Early Earth history is divided into epochs based on a combination of speculation and the age of rocks, meteorites, moon craters, various isotopes, and the sediment fossils. The 600-million-year Hadean Eon ended 3.9 billion years ago after an event known as the "late heavy bombardment." This marks the beginning of the Arch*ean era. Hard evidence for the Hadean consists of Earth forming a little over 4.5 billion years ago, bombardment of the Moon by asteroids 3.9 billion years ago, and the 3.8-billion-year-old chemical evidence of life. That's it. Scientists have drawn a lot of conclusions about the Hadean Eon that are weakly circumstantial and quite susceptible to interpretation.

But tracking the origins of Earth is a rapidly evolving detective story with geologists playing the role of Miss Marple or Hercule Poirot.

Let's begin with the scenario as most knowledgeable scientists understood it in August 2004 when we were in Kamchatka. Stetter had just stated that he thought life was about 4 billion years old and that it should be traced back through the volcanic vents and hyperthermophiles. This was entirely reasonable. In the early 1970s, geologists led by Vic McGregor, a New Zealander who had lived in Greenland for 30 years, were exploring an outcropping of very old rocks known as Isua Supracrustal Formation in West Greenland. McGregor sent samples of the rock to a laboratory headed by Stephen Moorbath to be dated based on lead isotopes. The age was determined to be 3.7 billion years old. In 1977, a second team led by A. Michard-Vitrac used a newer and more accurate technology to date rocks of the Greenland formation at 3.770 billion years old, rounded to 3.8 billion years. The newer date was determined by measuring isotope ratios of uranium decaying to lead from rare zircon crystals found in the rock. Zircon crystals are crucial to this detective story.

In 1979, Manfred Schidlowski discovered the first possible chemical evidence of life in rocks from the Greenland formation by analysis of carbon isotope ratios. In 1996, geologists Stephen Mojzsis at the University of Colorado and Clark R. L. Friend at Oxford University verified Schidlowski's discovery and dated the associated rocks at 3.8 billion years old. The low carbon isotope ratios from these ancient meta-sediments are associated with the metabolisms of microorganisms. Redox chemistries of archaea and primary-producing bacteria alter specific isotopes of different elements, including carbon and sulfur, which are main-menu items for archaea and bacteria today and presumably then. This metabolic evidence of life 3.8 billion years ago has held up pretty well. The Isua Supracrustal Formation results have been verified by several groups and are not under question.

In 1998, a team of Canadian geologists discovered an ancient outcropping called the Acasta Gneiss Complex near Great Slave Lake in the Northwest Territories of Canada. These rocks were dated at 4.03 billion years. The Acasta Gneiss Complex is still the oldest known actual rock on Earth.

With rocks dated at 4 billion years, scientists assumed this is when Earth formed a crust, thereafter making warm little ponds and oceans possible. Because of the 3.9-billion-year-old moon craters, everyone was certain young Earth experienced a violent childhood, whacked repeatedly during the first 600 million years by large objects called planetesimals, asteroids, and comets. The long bombardment is an assumption. Scientists assumed the repeated beatings kept Earth's surface as a molten magma sea and extremely hot until 4 billion years ago. Not much of an atmosphere could have existed. Molten Earth could not have developed a core until it cooled. With no core to increase gravity, hydrogen and all but the heaviest gases would have escaped, pfssssst, into space. Life could not have been possible until 4 billion years because of the intense heat, lack of an atmosphere, and arrested development caused by continual beatings.

Mythic Hadean Earth is depicted with many volcanoes and a heavily cratered surface. With hard evidence of crustal rock 4 billion years old, scientists assume Earth had then cooled enough to form a crust and a core. An atmosphere would then have developed from the degassing of the crust. Maybe the atmosphere was mostly carbon dioxide, maybe methane.

Earth's magma held a great reservoir of water molecules, which escaped as vapor at some point before 3.8 billion years ago along with a great reservoir of carbon dioxide. Volcanoes and fumaroles would have added a lot of sulfur. At some point between 4 billion and 3.8 billion years, the vapors condensed and it rained, rained, rained, further cooling the hot crust, sssssss, and the oceans formed. Without oxygen in the atmosphere the oceans would have been green for sure.

Scientists presumed that life must have arisen in hydrothermal vents or little warm ponds. Panspermia theorists suggest the building blocks for life, or life itself, were delivered to Earth by comets, which contain ice, organic compounds, and debris from the early solar system. Maybe comet soup was cooked in a pond, or maybe the raw organic ingredients of life were brewed in vents and stirred by violent bombardments. Lightning may or may not have been involved. Life, wherever it arose, did so spontaneously.

Some scientists suggest that, just as life was catching on 3.9 billion years ago, the late heavy bombardment might have wiped the slate clean, causing life to start over. This wipe-out-life-and-start-over business is probably wrong.

Bombardment by asteroids between 3.9 billion years and 3.84 billion years is well documented on the Moon. Barbara Cohen of the University of Hawaii, and David Kring of the University of Arizona, published definitive studies of this in 2000 in *Science.* In 2002, Cohen and Kring extrapolated from lunar impacts that Earth was hit by at least 22,000 objects, creating craters with diameters greater than 12 miles. They speculated the bombardment created 40 impact basins with diameters of about 600 miles, and several gargantuan craters more than 3,000 miles across: "one exceeding the dimensions of Australia, Europe, Antarctica or South America. The thousands of impacts occurred in a very short period of time, potentially producing globally significant environmental changes at an average rate of once per one hundred years." Cohen and Kring suggested the impacts created hydrothermal vent systems and led to the origin of life around 3.9 to 3.8 billion years ago. No physical evidence exists on Earth of the impacts that Cohen and Kring suggested. Impacts probably happened, but to what extent is not known.

The Acasta Gneiss Complex represents the oldest remnant of Earth's crust. The problem with finding actual rocks and especially fossils any older is that plate tectonics is constantly recycling rock. Igneous rock is crustal rock—the super-hard granites. The granite we mine today is relatively new. Metamorphic rock began as crust but was heated and compressed, destroying microscopic cell membranes and thus fossils. This recycling of the crust severely limits finding physical clues of first life on Earth. The impact craters of the heavy bombardment on Earth, however many existed, have vanished as old crust is made new again by tectonic forces. If massive craters have been remolded by tectonics since 3.9 billion years ago, the likelihood of finding fossils older than 4 billion years is slim.

But don't despair. Geologists have been dating rocks since at least 1911, and they are getting better at their methods all the time. Zircon crystal dating began in the 1950s. Zircons are a perfect match. Zircons are abundant in the crust and appear to be the oldest mineral on Earth; time capsules that contain an incredible amount of ancient history, said John Valley, president of The Mineralogical Society, and professor of geology at the University of Wisconsin.[1]

A typical zircon is the size of the period at the end of this sentence →.

Inside is a record of when Earth's crust formed, when the oceans appeared, and clues about baby Earth's atmosphere. Zircons may not reveal direct evidence of life, but dating various isotopes such as uranium, lead, and oxygen demonstrates when conditions for life arose. What zircons reveal is completely different from the Hadean picture.

Zircons are the product of granite formation. Valley said that as granite erodes, crystals are liberated and carried in dust and spread to other sediments and sand. Most ancient sediments have been buried and metamorphosed so that the crystals are now trapped in metamorphic rocks, which is old crust recycled and reformed into younger rocks. Crystals liberated from the original planet granites are elusive, but they have been found not far from the oldest known rocks in Western Australia, Greenland, South Africa, and Western Canada. In 1999, Valley and some of his students went zircon hunting in Australia.

"To recover less than a thimbleful of zircons, my colleagues and I collected hundreds of kilograms of rock from these remote outcrops and hauled them back to the laboratory for crushing and sorting, similar to searching for a few special grains of sand on a beach," Valley wrote in an article he authored for *Scientific American.*

Ordinarily this would be forced labor for graduate students, but zircons are so precious Valley and several other colleagues got their hands dirty too.

"Zircons are exceptionally robust and retentive. The most ancient grains were removed from unknown parent rocks, transported as wind-blown dust and river mud, and deposited as detrital grains in sedimentary rocks. They carry chemical clues to their origin in the form of mineral inclusions, trace elements, growth zoning, and isotopes of uranium, lead, oxygen, and hafnium," Valley wrote in his October 2005 article.

Valley's group found the oldest known zircon crystals at the Narryer Gneiss Terrane in Western Australia. (See Plate 7.) In 2000, they reported that the crystals are 4.4 billion years old, created during the formation of igneous rock, primarily granite. It took a few years for the significance to sink in and for a more complete picture to develop. The zircons had to have been created when molten magma began cooling and forming the first crust. This means Earth had a crust 4.4 billion years ago.

According to Valley, dating specific parts of a crystal first became possible in the early 1980s. William Compston and a team at the Australian National University in Canberra invented a special ion microprobe, which is a huge instrument. Compston's group called it the Sensitive High-Resolution Ion Micro Probe—SHRIMP. The instrument fires ion beams at the crystal with such precision it can "blast a small number of atoms off any targeted part of a zircon's surface," Valley said. A mass spectrometer measures the composition of the blasted atoms. Compston's group dated the first zircons using the ion probe in 1986 with Robert Pidgeon, Simon Wilde, and John Baxter. In 1999, Valley asked Wilde if they could use the instrument to perform additional isotope dating of zircons that his team was analyzing for oxygen isotope ratios. The 4.4-billion-year date was a shocker.

"No one had suspected that crystals this old might still exist on Earth. This meant the Earth's crust actually began forming as early as 4.4 billion years ago," Valley said.

Since then hundreds of zircons from the oldest known rock formations have been dated at ages ranging from 4 billion years to 4.4 billion years. The data also suggest Australia was one of the first continents. Australia rocks!

It appears the real Hadean period lasted only 100 million years. Further zircon crystal evidence showed the conditions for life as we currently understand it were in place no later than 4.2 billion years and possibly as early as 4.4 billion years ago.

"If the carbon isotope record in metamorphic rocks is correct, then the emergence of life was before 3.8 billion years. In fact, all of the essential ingredients for life were assembled on Earth as soon as near-surface waters cooled enough for DNA to be stable about 4.2 billion years ago," Valley said. "If life emerged on Earth or was delivered from space at this time, its main challenge would have been possible annihilation during the late heavy bombardment. Survival chances would have been enhanced if primitive microbes, like archaea, were capable of subsisting underground in the absence of sunlight. Alternatively, the earliest life on Earth may have evolved and become extinct many times before 3.85 billion years ago and the present inhabitants of Earth may be descended from the first continuously successful life, but not the first life."

Valley and an international team of five leading geologists put together the latest, greatest scenario of Earth's first billion years by summarizing thousands of published studies. Their scenario was published in the August 2006 issue of The Mineralogical Society of America's journal *Elements.* Much of the newest information is based on the zircon time capsules. With a richer understanding provided by their scenario, let's return to the beginning.

We know that meteorites formed 4.567 billion years ago based on mass spectrometry of isotopes contained in stardust and trapped in primitive meteorites. Some of the stardust belongs to our own sun and some of it belongs to a different system. Strong circumstantial evidence tells us these meteorites formed quite soon after our star was born. The meteorites, Earth, and the other planets formed at about the same time from a disk of gas and dust that immediately formed around the new sun. We know about the disk based on more than 200 years of mathematics and direct observations of other stars.

Our sun appears to have risen like the Phoenix from the collapse of a portion of a molecular cloud of gas and dust similar to the Eagle or Orion nebulae, according to Alex Halliday of the University of Oxford, and one of Valley's co-authors of the scenario study. Some of the stardust from that molecular parent cloud has been isolated from ancient meteorites.

Isotopes of elements in the stardust "provide fingerprints of the stars that preceded our sun," Halliday said. "Why this cloud of stellar debris collapsed to form our sun has been uncertain, but recent discoveries provide support for the theory that a shock wave from a supernova explosion provided the trigger."

The planets formed very soon after the sun. Numerous sophisticated and clever experiments show how the new sun would vacuum up fine dust and gas while bigger chunks of stuff turned into clods of matter and those clods stuck to each other by some type of "planetary glue" and those clods stuck to bigger ones and finally formed the terrestrial planets. Sometimes the clods, known as planetesimals or planetary embryos, would collide and crumble, only to reform into bigger terrestrial objects as they whipped around the sun.

Halliday said scientists agree planet formation occurred in four mechanistically distinct stages:

1. Settling of circumstellar dust to the mid-plane of the disk.

2. Growth of planetesimals of the order of 1 kilometer (0.62 miles) in size.

3. Runaway growth of planetary embryos of the order of 1,000 kilometers (620 miles) in diameter.

4. Growth of larger objects through late-stage collisions of the planetesimals.

No one is sure what sticky material helped the rocky planetesimals aggregate. Scientists think the "missing glue" was electrically charged and sticky. Sounds like space slime to me: organically rich, electrically charged plasma, cooked by thermonuclear events just prior to the first sunrise. Leftovers of the accretion disk, which include water, form comets, and tell us more about how space goo may have aided terrestrial planet formation.

After studies of the Hale-Bopp comet in 1999, NASA issued a press release from scientists who analyzed part of its composition with an infrared spectrometer. "Comets are interesting because they are frozen relics from the formation of our solar system, and by studying them, we can learn more about how we got here," said Dr. Michael DiSanti of Catholic University and NASA's Goddard Space Flight Center. "Our observations of Hale-Bopp indicate that comets now in the distant Oort cloud were originally part of the solar system's ancient protoplanetary disk. It was thought that comets could have formed in the cold, dense cloud of gas and dust that existed before the protoplanetary disk formed. However, if this were so, we would have seen even more carbon monoxide emission from Hale-Bopp. The amount of carbon monoxide ice compared to water (12 percent) indicates that these comets formed somewhere between the orbits of Jupiter and Neptune. We hope to learn more about what was going on when the giant planets formed by investigating the chemistry of this comet." Hale-Bopp was releasing carbon monoxide equivalent to the emissions of 5.5 billion cars per day.

That was a start. But no evidence of space slime. NASA's Deep Impact probe revealed more when it rammed comet Tempel 1 on July

4, 2005. The impact vaporized comet matter that allowed astronomers a crisp analysis of its composition. The imaging spectrometer on NASA's orbiting Spitzer Space Telescope provided the best tool for this job. The Spitzer team was led by Carey Lisse of the Johns Hopkins University Applied Physics Laboratory in Laurel, Maryland.

"Spitzer's spectral observations of the impact at Tempel 1 not only gave us a much better understanding of a comet's makeup, but we now know more about the environment in the solar system at the time this comet was formed," Lisse said in a press release. "Astronomers spotted the signatures of solid chemicals never seen before in comets, such as carbonates (chalk) and smectite (clay), metal sulfides (like fool's gold), and carbon-containing molecules called polycyclic aromatic hydrocarbons, found in barbecue grills or automobile exhaust on Earth."

Clay and carbonates were surprises. Liquid water is needed to make clay. Also surprising was the superabundance of crystalline silicates, material formed only at very high temperatures closer to the sun inside the orbit of Mercury. If comets contain clay and carbonates, so must the material in the accretion disk. We've got goo, and life-giving organics generated at the birth of the sun. According to Lisse, who published the results in *Science*, "In the same body, you have material formed in the inner solar system, where water can be liquid, and frozen material from out by Uranus and Neptune. Except for the lightest elements, the total abundances of atoms in the comet are practically the same as makes up the Sun. It implies there was a great deal of churning in the primordial solar system, with high- and low-temperature materials mixing over great distances."

Lisse reported: "The mixtures of materials seen in the ejecta [from the comet impact] are usually found in very different environments: highly volatile organic ices; clays and carbonates that form in aqueous environments; and highly crystalline silicates formed at temperatures exceeding 1000 K (1,340 degrees F). These results have implications for the structure and dynamics of the proto-solar nebula 4.5 billion years ago."

The gas cloud that collapsed to form the sun must have contained all of the raw materials for the sun and the accretion disk. Interestingly, some of the constituents of gas clouds similar to our own are sugars that form ribose, the backbone of RNA. What happens to the matter in the proto-solar nebula when it transforms into a hydrogen-burning star? It is obviously shaken, cooked, stirred, and converted into every particle of the solar system that becomes planets, forms chemistries, and gives life.

John Chambers of the Department of Terrestrial Magnetism at the Carnegie Institution of Washington, D.C., has conducted some of the most comprehensive studies of planetary formation. His worked was referenced by Valley's group.

"Water and other life-giving volatile materials are thought to have originally accreted in planetesimals located beyond 1 astronomical unit from the Sun in the early solar nebula (1 AU = distance from the Earth to the Sun). These small bodies were subsequently driven toward the inner solar system by the gravitational perturbations from Jupiter and Saturn," according to Chambers. "The evolution of the terrestrial planets and the asteroid belt was heavily dependent on the orbital characteristics of the giant planets. Our group demonstrated that the amount of volatiles present was affected by the timing of giant-planet formation."

The accretion disk, generated with everything necessary for life at the birth of the sun, seems to be more of a primordial stew than any hypothetical pond on Earth. The early solar system was an extremely energetic environment. I'm not sure we've considered all the possible opportunities under which various metabolisms would spontaneously arise. But in Earth's case, things got off to a much faster start than imagined.

Within 30 million to 55 million years of Earth's formation, we collided with another object roughly the size of Mars. This impact appears to have completely melted the spanking new Earth. The smaller object vaporized and the silicate dust left over became the Moon. Robin M. Canup of the Southwest Research Institute, Department of Space Studies, Boulder, Colorado, conducted 100 potential Moon-forming impact simulations based on Earth having accumulated about 80% of its current mass. She published the work in a report *Simulations of a Late Lunar-Forming Impact*, which is the coup de grace to remaining questions of the timing and events of lunar formation and its effects on Earth. Canup and Kevin Righter, senior research associate at the Lunar and Planetary Laboratory, University of Arizona, edited *Origin of the Earth and Moon*, outlining in 2000 the most probable scientifically based scenarios. Canup and others have subsequently published a number of refinements.

"The putative impactor planet, sometimes named Theia (the

mother of Selene, who was the goddess of the Moon), struck the proto-Earth with a glancing blow, accounting for the high angular momentum of the Earth–Moon system," Halliday added.

The Moon's formation so early plays a key role in dispelling some of the Hadean myth. Lunar studies reveal little evidence of significant bombardments from the time of its formation until the late heavy bombardment 3.9 billion years ago. This suggests claims about our violent childhood are exaggerated. Not a whole lot of whacking was going on until the late bombardment.

This means young Earth had a more stable youth and was molten for no more than 100 million years. After recovering from the collision with Theia, our melted planet would have been covered almost instantly in a deep ocean of molten magma. Christian Koeberl of the University of Vienna, Austria, said heavier iron elements likely sank rapidly to the center of Earth via gravity and formed the Earth's core much sooner than the violent Hadean theory allowed. The early core means Earth could have developed a much thicker atmosphere within the first 100 million years of its birth and would have begun cooling right away.

"The magma ocean may have been shorter lived than previously thought, and differentiation and recycling [of crust] could have started shortly after the Moon-forming impact," said Koeberl.

The impact with Theia also likely vaporized some of Earth's mantle. Straightaway, Earth's atmosphere thickened with massive clouds of silica from Theia and our own mantle, which was as hot as 4,200 degrees F. The silica clouds probably completely swaddled baby Earth.

"For a thousand years the silicate clouds defined the visible face of the planet. The new Earth might have looked something like a small star or a fiery Jupiter wrapped in incandescent clouds. Silicates condensed and rained out at a rate of about a meter a day. Mixed into the atmosphere, at first as relatively minor constituents but becoming increasingly prominent as the silicates fell out, were the volatiles. Because convective cooling requires that every parcel be brought to the cloud tops to cool, the mantle should have largely degassed, with the notable exception of water, which remained mostly in the molten mantle and which degassed as the mantle froze," wrote Kevin Zahnle of the NASA Ames Research Center in California.

A hot atmosphere quickly formed of carbon dioxide, carbon oxide, water vapor, hydrogen, nitrogen, and the six noble gases: helium, neon, argon, krypton, xenon, and radon, according to Zahnle's report, *The Earth's Earliest Atmosphere.*

"The Moon-forming impact may or may not have expelled a significant fraction of Earth's pre-existing volatiles, and Earth may or may not have had abundant volatiles to lose. A primary hydrogen atmosphere, because of its low mean molecular weight, would have readily escaped. But a secondary atmosphere would have to be pushed off. It is generally agreed that the volatiles on the side of Earth that was hit were lost, but it is an open question how volatiles on the other side could be lost. Recent theory suggests that the answer depends on whether there had been a deep liquid-water ocean on the surface. A thin atmosphere above a thick water ocean can be expelled. Otherwise, the atmosphere is retained. One notes that water is retained in either event. The view taken here is that the planet that became Earth was water rich," Zahnle wrote.

Zahnle was beating around the bush a bit. With crustal formation at 4.4 billion years, the planet began degassing like crazy. Valley and coauthors concluded from additional zircon crystal data that the atmosphere was thicker quicker and Earth cooled in a hurry.

One version of the Hadean hypothesis suggested Earth's water was delivered by comets. Valley and his colleagues argued that comets could not have delivered enough water to account for the ocean and stressed again that the accretionary material of Earth came ready-mixed with water, providing an ample reservoir in the mantle and crust. As water steamed out of a cooling surface, it would have condensed as clouds and returned as rain to create emerald oceans with little oxygen and lots of CO_2. "The water was first condensed as ice, either locally in the planetesimals from which the bulk of Earth was made, or in more distant planetesimals scattered from what is now the asteroid belt, or in comets. The likelihood that any known source could deliver an ocean of water to Earth after the Moon-forming impact is demonstrably small," Zahnle wrote.

The aftermath of the Moon-forming impact left Earth with a hot, carbon dioxide–rich steamy atmosphere. Oceans condensed from the steam in a brief 200 million years and for some 10 to 100 million years the surface was a mild 450 degrees F or so. (Clouds would have been cooler, within temperature ranges for early life.) The upper temperature for modern hyperthermophiles is 250 degrees F. Zahnle and the others concluded that life may not have been present on the surface.

"Thereafter a lifeless Earth, heated only by the dim light of the young Sun, would have evolved into a bitterly cold ice world. The cooling trend was frequently interrupted by volcanic- or impact-induced thaws," Zahnle wrote.

The sun was about 30% dimmer shortly after its birth than it is today. Zahnle said that Earth may have become bitterly cold by 4.4 billion years ago. For Earth to have been more temperate, the planet would have had to generate more heat or the atmosphere would have had to be filled with greenhouse gases. Based on the latest models and physical evidence, Earth did not radiate enough heat after 4.4 billion years ago, aside from being hit periodically by asteroids. He said the only good candidates for greenhouse gases were carbon dioxide and methane.

"Methane would be a candidate if there were reducing agents or catalysts to generate it from CO_2 and H_2O. On Earth today methane is mostly of biological origin. Methane is a good candidate for keeping Earth warm once it teemed with life, but it is not clear that there was a big enough source of it when Earth was lifeless," Zahnle said.

If life formed in the clouds, it could have provided the catalysts for reducing methane. Just a thought.

The new scenario envisioned an icy planet with periodic 100-year-long summers caused by impacts. This model is consistent with the extrapolations made by Cohen and Kring, except that Valley's group suggested that impacts were less frequent. Additional zircon dating strongly suggests Earth had a fully developed CO_2 atmosphere and deep oceans by 4.2 billion years ago.

"Thus, liquid water seems to have been present on the surface of Earth early on, and granitic (not just basaltic) pockets of continental crust were present," Koeberl said. "We can infer that the frequency of meteorite impacts during the time span between 4.4 billion and 4 billion years ago may have been less than previously thought."

In 2005, Bruce Watson, a geochemist at Rensselaer Polytechnic Institute in Troy, New York, and Mark Harrison of the Australian National University in Canberra analyzed oxygen isotope data preserved in zircon crystals to demonstrate the presence of water. They determined that zircon crystals form at a temperature of 1,292 degrees F, which is actually relatively cool. "The temperatures substantiate the existence of wet, minimum-melting conditions within 200 million years of solar system formation. They further suggest that Earth had settled into a pattern of crust formation, erosion, and sediment recycling as early as 4.35 billion years ago," Watson and Harrison concluded in a study they reported on in *Science.*

Long-time science reporter Robert Boyd of the Knight Ridder Newspapers attended a NASA Astrobiology Institute conference in April 2005 where the data were discussed. Harrison told Boyd in an interview, "We don't know when life began on Earth. But it could have emerged as early as 4.3 billion years ago. Within 200 million years of the Earth's formation, all of the conditions for life on Earth appear to have been met."

Stephen Mojzsis, who chaired the NASA astrobiology conference, told Boyd: "The stage was set 4.3 billion years ago for life to emerge on Earth. There was probably already in place an atmosphere, an ocean, and a stable crust within about 200 million years of the Earth's formation. Water was gushing out of the Earth."

If Cohen and Kring are correct in asserting that impacts could generate hydrothermal activity, then vents could have been present anytime between 4.4 billion and 4.2 billion years ago. Earth's developing crust and tectonics also would have had hundreds or thousands of hot spots similar to the Valley of the Geysers. Life could have been flourishing by 4.2 billion years ago. If it existed in hydrothermal systems it would most likely have resembled modern archaea. Asteroid impacts on the cold icy Earth from about 4.4 to 4.3 billion years ago would have warmed the atmosphere and generated hydrothermal processes.

Variations in oxygen isotope composition of zircons with ages of 4.2 billion years indicate that they grew in magma that cooled and formed part of the crust. These studies were made by Aaron Cavosie, assistant professor at the University of Puerto Rico. The isotopes show the granite at this time was exposed to weathering, and chemical and structural changes at low temperatures.

Some scientists have suggested that the late heavy bombardment sterilized Earth, forcing life, if it existed, to start over again. But newer calculations show the impact energies associated with the late heavy bombardment at 3.9 billion years ago were not high enough to vaporize the oceans or sterilize the planet. Scientists believed Earth was subjected to a 600-million-year-long rain of fire based on assumptions of a debris-filled solar system and the evidence of the heavy bombardment 3.9 billion years ago. But Chambers said studies of impact craters on the Moon show the late heavy bombardment was an aberration. Chambers had an intriguing proposal for this: the early solar system contained *five* terrestrial planets inside the asteroid belt instead of four.

"I believe that the missing fifth planet was in an unstable orbit between Mars and the asteroid belt and was ejected by 600 million years of gravitational perturbations induced by the other planets. I am proposing that the missing planet's exodus disrupted asteroid fields, creating an increase in lunar impacts," Chambers said.

Baby! A fifth planet, dropping out just when the solar system was really rocking. The Beatles began with five members too. I like Chamber's fifth planet hypothesis. That's page 1 material. Please give us more data.

The geologists have established a new evidence-based scenario for early Earth and dispelled much of the Hadean mythology. At this writing, the earliest evidence for life remains in the carbon isotope record of ocean sediments at 3.8 billion years. But we know the conditions for life were present as early as 4.4 billion years ago and no later than 4.2 billion years.

Now let's return to the problems of fossils. Locating fossils more than 200 million years old is tough enough. Older than 2 billion years is downright hard. The fossil history of Earth is divided broadly into two eras—the Cambrian and Precambrian. It is very simple. Pre-Cambrian = next to nothing. Cambrian = just about everything. The dividing line is at about 560 million years ago with the "Cambrian Explosion." This is when remarkable forms of multi-cellular life appeared on Earth en masse, as if from some interplanetary ark.

Darwin pointed out this fossil record mystery in 1859: "If the theory [evolution] be true it is indisputable that before the lowest Cambrian stratum was deposited . . . the world swarmed with living creatures. [Yet] to the question why we do not find rich fossiliferous deposits belonging to these earliest periods . . . I can give no satisfactory answer. The case at present must remain inexplicable."

The search for fossils of microorganisms has been focused on stromatolites, which appeared about 3.5 billion years ago. Stromatolite layers can vary from less than an inch to more than three feet thick. Not everyone agrees, but the majority of scientists are convinced that stromatolites are fossils of ancient microbial mats.

The Paleontological Society provides a nice background and description on stromatolites: "A major breakthrough came in the mid-

1960s when it was recognized that thinly layered centimeter- to meter-sized rock masses known as stromatolites can be richly fossil-bearing. To the stratified microbial menagerie that builds such structures, stromatolites are rather like high-rise apartment houses."

The "penthouse" upper floors are occupied by light-using photosynthetic cyanobacteria and other microbes that respire oxygen. Beneath the penthouses, also on upper floors, are the photosynthetic bacteria that rely on dim light that seeps through the cyanobacteria layer, and the switch-hitters (facultative aerobes) that use oxygen when available or revert to anaerobic metabolism when necessary. "The lower floors and basement are inhabited by strict anaerobes, bacteria for which oxygen is a deadly poison. Mineral debris accumulates on the uppermost layer, and the whole structure can become lithified. The shape of a stromatolite reflects its environment—flat-layered forms are typical of quiet water settings whereas mound-shaped and columnar types occur in regions of turbulent wave action. Stromatolites are very common in Precambrian limestones, rocks made up of the mineral calcite, but the best microscopic fossils occur in stromatolites that have been petrified by fine grained quartz, silicified like the logs of a fossil forest." (See Plate xx.)

In 1986, geologist William Schopf was searching the "Pilbara Block" region of Western Australia, a wonderland of well-preserved stromatolites. In this ancient landscape, stromatolite features stretch for miles with all shapes and sizes. Some look like inverted egg cartons, others are cones, and some are columns. Schopf reported in *Science* three types of formations that showed evidence of microscopic fossils dating back to 3.5 billion years. He continued working in the region and put himself on the map as well as the firing line of critics with the publication of a whopper of a paper on April 30, 1993. Schopf's summary conclusion: "Eleven taxa (including eight heretofore undescribed species) of cellularly preserved filamentous microbes, among the oldest fossils known, have been discovered in a bedded chert unit of the Early Archaean Apex Basalt of northwestern Western Australia. This prokaryotic assemblage establishes that trichomic cyanobacterium-like microorganisms were extant and morphologically diverse at least as early as [3.465 billion] years ago and suggests that oxygen-producing photoautotrophy may have already evolved by this early stage in biotic history."

No one knows whether these organisms were bacteria or archaea, or something in between. Some archaea and bacteria were capable of oxygen respiration at very low levels. But it is doubtful that the penthouse floors contained any oxygen-using cyanobacteria 3.5 billion years ago. The atmosphere consisted primarily of CO_2 until 2.7 billion years ago when oxygen levels began rising. The fact that modern microbial mats consist of consortiums with anaerobic species at the bottom, metabolic switch-hitters in the middle, and oxygen-using cyanobacteria on top suggests the order could have been flipped during the CO_2-dominated Archean Empire. As oxygen increased in the atmosphere, cyanobacteria could have migrated to the penthouses.

In 1984, Alec Trendall, a former Director of the Geological Survey of western Australia, discovered one of the first rich fossil beds of stromatolites in the Pilbara craton, 300 miles north of where Valley obtained his samples in the Yilgarn Craton, a huge stretch of ancient continental crust that accounts for most of western Australia's land mass. Hans Hofmann, Kath Grey, and Arthur Hickman returned and found an even richer cluster in the 3.45-billion-year-old sediments of the eastern Pilbara, according to the Australian Centre for Astrobiology. Volcanic rocks associated with the stromatolites were dated using uranium/lead isotopes from zircons, which give very precise ages at 3.4 billion years for the Trendall formation, 3.490 billion years for the nearby "North Pole" stromatolites, and 3.460 billion years for the Chinaman Creek stromatolites.

"The stromatolites have a variety of shapes that include steep-sided cones, branching columns, and tall, single columns, and there are flat laminae between the positive features," Hofmann reported.

No actual microfossils have been discovered so far, but the microstructure of the stromatolites is consistent with formation by bacterial mats. Trendall and his colleagues said the Australian stromatolite-like structures are unlikely to have formed through deformation or by chemical processes because their sizes and shapes are too complex. None of the features observed are typical of strictly geological formation caused by erosion and chemical precipitation of minerals. Chemically precipitated layers would have a regular thickness, whereas the mats have layers that are thicker than others, with the thickest at the bottom or center of the structures. "The laminae thicken over the cone apices and thin on the flanks. In most of their features, the structures resemble stromatolites of undoubted biogenic origin found in younger rocks, and it is difficult to explain their formation by any other means than a biological one," Trendall said.

Interpreting stromatolites is careful business. You must do everything possible to support a claim, especially if it is radical, like a new domain of life or the most ancient microbial fossils on the planet. Scientists rarely hesitate to point out someone else's mistakes. Schopf learned this nearly a decade after his 1993 *Science* paper. Geologist Martin Brasier of St. Edmund Hall, Oxford, and his team reanalyzed Schopf's data and in 2002 published a critique concluding the stromatolites were not microfossils, but merely natural accumulations of organic material and minerals glommed together at an ancient hydrothermal vent system. Brasier may be right about the particular fossils that Schopf analyzed in 1993, but most scientists in the field have no doubt that other 3.5-billion-year-old stromatolites are microbial in origin.

According to Schopf:

> In recent decades, rules for accepting Precambrian microfossil-like objects as bona fide fossils have become well established. Such objects must be demonstrably biogenic, and must also be indigenous to and have formed at the same time as rocks of known provenance and well-defined Precambrian age. To address biogenicity—the most difficult of these criteria to satisfy—a nested suite of seven traits has been proposed; characteristics identifying spheroidal or filamentous fossil microbes, the predominant Precambrian morphotypes, have been specified; and the usefulness of such traits has been documented. A prime indicator of the biological origin of microscopic fossil-like objects is the co-occurrence of distinctive biological morphology and geochemically altered remnants of biological chemistry. Thus, chemical data demonstrating that populations of 'cellular microfossils' are composed of carbonaceous matter would be consistent with a biogenic interpretation.

Fossils of microbial mats meeting the strict criteria include:

- Stratiform and conical stromatolites dated at 2.985 billion years from the Insuzi Group formations in South Africa.
- Laterally linked, low-relief, stratiform to domical stromatolitic mats dated at 3.245 billion years from the Fig Tree Group formations of South Africa.
- Stratiform microbial mats dated at 3.320 billion years from the Kromberg formation of South Africa.
- Conical stromatolites dated at 3.388 billion years from the Strelley Pool Chert of Western Australia.
- Domed and stratiform stromatolites dated at 3.496 billion years from the Dresser Formation of Western Australia.

In June 2006, Abigail Allwood of the Australian Centre for Astrobiology reported another colossal find from the Pilbara formations. Allwood's team studied a six-mile alien-like landscape covered in stromatolites of different shapes and sizes and reported their results in *Nature.*

Many of the structures are small columns. Others resemble an upside-down egg carton—miles of them in rows. The team identified seven specific shapes and types of formations that are biological in origin. Allwood and her colleagues argued that random geological processes could not possibly create such a large cluster of diverse formations as those at Pilbara. These are dated at 3.4 billion years old. The team compared the shapes and features of the stromatolites with those of structures in the same area known to be purely mineral deposits and concluded microorganisms built the stromatolites. Can you imagine six miles of ancient microbial mats stretching beneath a shallow sea? (See Plate 8.)

In a press release issued by the Australian Centre for Astrobiology Allwood said:

> This is the first discovery of a large, complex, ecosystem-scale remnant of the early biosphere, an entire fossilised community of diverse stromatolites, spanning rock outcrops more than 10 kilometres in length. Due to the excellent preservation and broad extent of the outcrops, we have been able to study the group of stromatolites and their associations with each other within their early environment. We are able to see how this early ecosystem grew and changed in response to different environmental conditions across a wide area over a period spanning 80 million years. This has shed light on the conditions that nurtured very early life, and we can use that to predict where life might have emerged on other planets. Our next big question is about the nature of the microorganisms that produced these structures. We believe that many different types of organisms may have coexisted at this time, so that we have not just some of the oldest evidence of life, but we also have the oldest evidence of biodiversity. One of the broader implications of this work is in helping us search for traces of life on Mars and other planets. If you're going to find anything on a planet like Mars it's going to be unlike anything we expect to see on our own planet today. It will likely be more like the type of micro-organisms that produced the early stromatolites. Therefore, the Strelley Pool Chert is an excellent case study for learning how to detect and identify signs of life on Mars. These fossils provide critical insights into the origins of life and are literally irreplaceable. It is essential that we recognise their importance and preserve them.

Science reporter Robin Williams conducted an interview with Allwood for his popular Australia ABC National Network program, *The Science Show*, when the study was released in June 2006. Allwood told Williams, "I think sort of the Holy Grail of these kinds of studies has been to determine the antiquity of life on Earth, which is a nice thing, and this is not necessarily evidence of the oldest life on Earth. I think life could have arisen much earlier than this, it looks like it was quite well established."

Genetic analysis of living stromatolites in Australia's Shark Bay was conducted by Norman Pace, Dominic Papineau, Jeffrey J. Walker, and Stephen J. Mojzsis, at the University of Colorado. Shark Bay, also in Western Australia, is a relatively rare underwater field that looks conspicuously similar to the fossil formations at Pilbara. Pace and his group applied 16s rRNA methods to obtain microbial ecology data on the living stromatolites. Here's what Pace's group reported in the *Journal of Applied and Environmental Microbiology* in August 2005:

> Stromatolites, organosedimentary structures formed by microbial activity, are found throughout the geological record and are important markers of biological history. More conspicuous in the past, stromatolites occur today in a few shallow marine environments, including Hamelin Pool in Shark Bay, Western Australia. Hamelin Pool stromatolites often have been considered contemporary analogs to ancient stromatolites, yet little is known about the microbial communities that build them. We used DNA-based molecular phylogenetic methods that do not require cultivation to study the microbial diversity of an irregular stromatolite and of the surface and interior of a domal stromatolite. . . . The communities were highly diverse and novel.

The library of sequences created from the stromatolites showed the mats were 90% bacterial and 10% archaeal. The most abundant sequences from the stromatolites were representative of unknown proteobacteria (28%), planctomycetes (17%), and actinobacteria (14%). Sequences representative of cyanobacteria, long considered to dominate these communities, comprised less than 5% of the population. Approximately 10% of the sequences were most closely related to those of proteobacterial anoxygenic phototrophs—the middle layers, most likely. (See Plate 8.)

"These results provide a framework for understanding the kinds of organisms that build contemporary stromatolites, their ecology, and their relevance to stromatolites preserved in the geological record," Pace and his team concluded.

Cyanobacteria start showing up clearly in the fossil record by 2.7 billion years ago and more commonly since 2.3 billion years ago. Scientists at first thought modern cyanobacteria were plants so they were formerly known as blue-green algae. They are bacteria capable of plant-like photosynthesis. You might know these microorganisms as pond scum. Before the cyanobacteria, life on Earth converted light into energy the way salt-loving archaea do so today with chemical photoreceptors.

The zircon isotopes reveal that 2.4 billion years ago the atmosphere began spiking with oxygen. The toxic transformation was complete by about 1.5 billion to 1.6 billion years ago. This is the "Oxygen Catastrophe" I mentioned in the previous chapter. Two factors appear to be in play in the Oxygen Catastrophe, which reversed the boundaries of the outer limits. The first factor may be due perhaps to the runaway success of anaerobic bacteria and archaea that liberated vast amounts of oxygen from the crust after consuming the carbon dioxide in the atmosphere and ocean like there was no tomorrow. They became so numerous and so successful that they depleted their natural resources and polluted their water and air beyond redemption with oxygen. The Archean Empire began to crumble. The microorganisms that could tolerate the poisonous O_2, the cyanobacteria, began to expand and dominate. Archaea were driven underground. Their zone of habitability narrowed. Eventually, the once dominant extremophiles would be regarded as life on the edge, hosting raves in deep dark holes, doing hydrogen and sniffing sulfur. The most likely candidates for early life would include anaerobic archaea and bacteria. Determining whether archaea came before bacteria, or vice versa, may be irrelevant and probably impossible.

Maybe life doesn't care what form it takes. More important than the structure of the first organism or organisms is metabolism. Follow the energy trail and you'll find the origin of life. Metabolism matters. Everything else is adaptive packaging. Whatever kinds of life arose first on Earth or anywhere had to obtain energy and use it. Astrobiologists may discover metabolisms we haven't thought of or recognized while microbiologists are certain to find new permutations as they learn more about microbial consortiums. Life is driven to use energy as efficiently as possible. The first metabolisms would have been just efficient enough.

"When people think of diversity, they usually think about different types of flowers and the different shapes of animals. Microbes don't display much variety of shapes when you see them under a microscope. So when we discuss microbial diversity what we're really talking about is metabolism. Life is much more than a single cell replicating itself. It's a combination of different chemistries working together," Short said.

Because of Darwin's suggestion, the mythic, long, hot, violent Hadean epic, and the focus on hyperthermophiles by biotechnology, most of us have taken for granted that life arose in a warm pond or a hydrothermal system. Scientists have driven it into our heads that life—thus metabolism—requires heat. But it doesn't.

Ricardo Cavicchioli has focused on the psychrophiles—the cold-adapted archaea. At the time of this writing he was preparing to depart in December 2006 for his first expedition to collect samples from the Antarctic. Cavicchioli is a microbiologist at the School of Biotechnology and Biomolecular Sciences, University of New South Wales, Sydney, Australia.[2]

He said:

> I will be going for the first time this summer, leaving in December. Despite working on archaea from Antarctic Lakes for about 11 years, I have resorted to living vicariously through others. While I had opportunity to go, the main reason I didn't is that I feel strongly that people should only step foot on the Antarctic continent if they have a very good reason to do so. It should not be a tourist destination, and even scientists can be tourists. Fortunately I now have a very good reason to go, which relates to having funding from Craig Venter and the Joint Genome Institute to perform genome sequencing of environmental samples and metagenomics taken from lakes in the Vestfold Hills. My decade of research, in concert with new technological developments in genomics, have come together to provide both the valid questions and the means to address the questions. The Ace Lake system will be our port of call on this trip, and I will finally have my opportunity to stand on the ice surface of the lake that for years I have only dreamed about, and dip below its surface to discover some of its hidden secrets that I have longed to discover. Amidst the frantic five-day window in which a team of four of us will work 24 hours a day to carefully take our samples, I expect it will be a reverent and personal experience, and one which I expect I will never forget! And yes, perhaps I'll be so engrossed that I will fall through the ice never to be seen again! It really will be living a dream for me.

I was surprised to learn that the largest proportion and greatest diversity of archaea exist in cold environments. We forget Earth is, or was, primarily a cold, icy planet. We just happen to be living during an interglacial period. Even so, about 75% of the Earth's biosphere is cold. This summer vacation we're on has only been under way for 12,000 or so years. Unless we've permanently altered the climate through global warming, in which case we will soon destroy human civilization, we can expect this long summer to end in another 10,000 years or sooner. Glacial periods on Earth tend to be much longer, easily lasting 80,000 years.

Based on the new scenario from Valley and his crew, early Earth was cold too. Life did not have to begin in a warm anything. According to a summary of research on cold-adapted archaea published May 2006 in *Nature Reviews* most psychrophilic microorganisms prefer temperatures at 14 to 32 degrees F. They metabolize nutrients in snow and ice at –4 degrees F, and possibly as low as –40 degrees F. Archaea survive at –50 degrees F. Archaea are found in abundance in permanently cold environments that never rise above freezing, such as alpine levels on mountains and at both the North and South Poles. They live in the deep ocean; cold, dark caves; and in the thin, cold, radioactive, upper atmosphere. Cold-adapted archaea are most abundant in the ocean, and they are increasingly found engaging in consortiums bizarre enough to rival the rusticles. Cold-adapted methanogens are especially collaborative with other archaea and bacteria. Methane-generating metabolisms are not as efficient as those of other archaea, so it's not uncommon to find cold-adapted methanogens coupling with sulfur-reducing or nitrogenic bacteria. We don't hear much about the methanogens living on cold methane seeps in the dark deep and in methane ice. *Methanococcoides burtonii* and *Methanogenium frigidum* reside at Ace Lake, Antarctica. Cold-adapted archaea have been recovered from Scan Bay, Alaska; Lake Soppen, Switzerland; and Gotland Deep in the Basaltic Sea. Permafrost methanogens have metabolisms that function at both freezing and thawing temperatures.

Cold-adapted archaea play a significant and yet underappreciated role in the global energy cycle. Methane is a potent greenhouse gas. Vast quantities of methane have been stored for eons in ice. When glaciers melt and permafrost evaporates, methane is released into the atmosphere.

According to Cavicchioli, microbial methanogenesis in the seafloor subsurface is responsible for the formation of methane-hydrate deposits, which have accumulated four to eight times the amount of carbon in all living organisms on Earth. A cycle of anaerobic methane turnover occurs periodically in the cold ocean floor, where not only methane production, but also anaerobic oxidation, is driven by members of archaea.

"Permafrost covers more than 25% of the Earth's land surface, and methanogens in permafrost have been estimated to contribute about 25% of the methane release from natural sources," Cavicchioli wrote. "Despite this global impact, little is known about the methanogens that produce the methane and how they function in the cold."

Do we want to be encouraging the growth of methanogens and CO_2-loving microbes? Microorganisms exert tremendous influence on the geochemistry of Earth's atmosphere. We might want to figure this out. Could we trigger a change in Earth's atmosphere back to its carbon dioxide/methane composition?

The Valley of the Geysers is buried in snow at least six months a year. I gazed at the bubbling mud and snow-covered ridges and felt an inexplicable joy at the raw power of microbial life. Every expedition into an extreme environment is a surrogate for what might exist beyond Earth. Yellowstone Lake and the polar icecaps are models for Europa. The Valley of the Geysers is a model for Jupiter's moon Io and early Earth. Cold methane seeps in the Gulf of Mexico are models for Titan. Parts of Yellowstone and the Atacama Desert of Chile are models for Mars. Atacama is the driest place on Earth. In 2003, scientists reported that the soils of the Atacama Desert were sterile. But a team from the University of Arizona (AU) dug deeper and found colonies of microbes, yet unidentified.

AU's Peter H. Smith, principal investigator for the Phoenix mission to Mars, said, "Scientists on the Phoenix Mission suspect that there are regions on Mars, arid like the Atacama Desert in Chile, that are conducive to microbial life. We will attempt an experiment similar . . . on Mars during the summer of 2008."

A study at Yellowstone at the Norris Geyser Basin also had implications for finding evidence of life on Mars. This one was conducted by Jeffrey Walker and Norman Pace of the University of Colorado. The NSF and NASA provided a grant in 2005 for a very clever idea. Walker and Pace discovered microbes living in the pores of rocks in the acidic basin with high concentrations of metals and silicates at about 95 degrees F. Microbial colonies in rocks are likely to be preserved eventually as fossils, Pace said. Geothermal environments may have existed on Mars and hold fossil records of previous Martian microbes.

"This is the first description of these microbial communities, which may be a good diagnostic indicator of past life on Mars because of their potential for fossil preservation," said Walker. "The prevalence of this type of microbial life in Yellowstone means that Martian rocks associated with former hydrothermal systems may be the best hope for finding evidence of past life there."

"The pores in the rocks where these creatures live have a pH value of one, which dissolves nails," Pace added. "This is another example that life can be robust in an environment most humans view as inhospitable."

The colonies were discovered in 2003 when Walker broke open a sandstone rock. Back at the lab, the team found a green band within the rock made by a new species of photosynthetic microbes that are among the most acid-tolerant photosynthetic organisms known. "Cyanidium" algae-type organisms made up about 26% of the microbes identified in the Norris Geyser Basin study. A new species of mycobacterium associated with tuberculosis and leprosy made up 37% of the population of species found in the basin, Walker said. These pathogens had never been identified in such extreme hydrothermal environments.

Of the relationship between the two types of organisms, Pace said, "It is pretty weird. It may well be a new type of lichen-like symbiosis. It resembles a lichen, but instead of being a symbiosis between a fungus and an alga, it seems to be an association of the mycobacterium with an alga."

In their *Nature* paper, the team wrote, "Remnants of these communities could serve as 'biosignatures' and provide important clues about ancient life associated with geothermal environments on Earth or elsewhere in the Solar System."

Some of the microscopic fossils found on Earth could be ancient microorganisms that might have lived on Mars. In the early days of the solar system, asteroid collisions could have nicked off pieces of Mars and sent them spiraling toward Earth. Scientists argue that a fair amount of exchange of rocky matter could have occurred between planets. When a meteor enters the atmosphere it literally creates a hole which allows ejected matter to bounce through unaltered and out into space. It's possible some promiscuous swapping of ancestral life forms took place between worlds similar in a grander scale to the promiscuous swapping of genes between early cells.

Dozens of excellent books have been written about the topic of astrobiology. David Grinspoon's *Lonely Planets: The Natural Philosophy of Alien Life* is a good one that will take you on a better ride than I could into space. I am going to stick with what microbes on Earth are suggesting about life up there. But for space junkies, the most startling single conclusion I came across is that there is no evidence of a minimum temperature for metabolism.

Buford Price of the University of California, Berkeley, and Todd Sowers of Pennsylvania State University, arrived at this conclusion after reviewing more than 60 research papers on microbes in very cold and inhospitable conditions. In some of the studies microorganisms were tortured at various temperatures with all manner of substances, deprived of nutrients to the brink of starvation, you name it. The papers also importantly covered *in situ* conditions in frigid ecosystems all over Earth. This work raises the prospects of life beyond Earth.

These are a few gems Price and Sowers described from studies of the lower temperature extremes:

- Concentrations of 200 to 5,000 bacterial cells per milliliter in surface snow and firn—which is compressed snow at the verge of becoming glacial ice—at the South Pole; 10–20% were members of the *Deinococcus-Thermus* group. These are different types of bacteria grouped together phylogenetically. *Deinococcus* are extremely radiation resistant.
- Airborne bacteria in clouds can grow and divide at ambient air temperatures as low as 15 degrees F with sufficient dissolved organic carbon in cloud droplets to support bacterial growth.
- Bacteria can synthesize molecules for their cells in ice at 5 degrees F, at a rate that is very low but sufficient to repair cells. It is possible cells remain metabolically active in water films on the surfaces of mineral grains in ice or in microscopically small liquid.
- Metabolic activity of sulfate reducers and methanogens in marine sediments living nearly 1,000 feet *below* the ocean bottom is greatly concentrated in relatively narrow zones together. They appear to be using each other's metabolites in a syntrophic relationship. Scientists concluded these organisms are adapted for extraordinarily low metabolic activity.

- Huge excesses of carbon dioxide 32 times higher than in the densest part of the atmosphere; methane levels 8 times higher; and nitrogen levels 240 times higher in ice cores at depths corresponding to ages of 14,000 years ago for carbon dioxide and nitrogen, and 15,400 years ago for methane. The gases were trapped in air in the ancient ice. The gases did not represent levels of ancient atmospheres. They were the products of the active metabolisms of microbes living in microscopic-size veins of water in the ice. These veins of liquid water can arise when certain compounds are concentrated.

The microbes colonizing ice appear to have two levels of metabolism: a survival metabolism in which they remain alive but become dormant until given nutrients or higher temperatures, and maintenance metabolism for sustained growth. Some organisms in permafrost appear to have protein repair enzymes that maintain active recycling of certain amino acids needed for cellular repair for at least 30,000 years.

"The extremely low expenditures of survival energy enable microbial communities in extreme environments to survive indefinitely," Price and Sowers reported. They said nitrifying bacteria with low but active metabolisms have been found encased in liquid veins at –40 degrees F in Vostok ice for more than 140,000 years. It takes about 108 years for carbon to turn over in the cells. Scientists also found microbes surviving in clay and shale for more than 100 years. Here is the best part:

> Our results disprove the view that the lowest temperature at which life is possible is –17 degrees C (about 1 degrees F) in an aqueous environment, as well as the remark that "the lowest temperature at which terrestrial and presumably Martian life can function is probably near –20 degrees C (–4 degrees F)." Our data show no evidence of a threshold or cutoff in metabolic rate at temperatures down to –40 degrees C. A cell resists freezing, due to the "structured" water in its cytoplasm. Ionic impurities prevent freezing of veins in ice and thin films in permafrost and permit transport of nutrient to and products from microbes. The absence of a threshold temperature for metabolism should encourage those interested in searches for life on cold extraterrestrial bodies such as Mars and Europa.

Life, moving so slowly as to appear frozen, dormant, or undetectable may survive the cold, icy, and cosmically radioactive conditions of outer space. The last night we spent at the Valley of the Geysers the sky was crystal clear, and the stars were as thick as clouds. Meteors streaked across the eastern sky. I've long been fascinated with the idea that life is everywhere, that stars are big incubators; solar systems are chemical gardens. I bet that life arose very early on Earth, initially in big clouds and then fell to the surface with the first raindrops, taking root wherever possible. My secret hunch is that life in some primitive simple form was spontaneously generated by the ignition of the "main sequence," when our sun began burning hydrogen and energy was transferred from the cosmos into our new solar system. In a flash, all the elements were spread in a sumptuous banquet: carbon, oxygen, hydrogen, nitrogen, phosphorus, everything needed for nucleic acids, membranes, proteins, and the components we consider essential to a cell.

Everything was delivered from the celestial kitchen though our brand new sun. Heat, stir, and add water. Allow to cool 100 million years.

8
MONEY FOR NOTHIN', BUGS FOR FREE

Averting the rise of the second Archean Empire.

My momma always said, "Life was like a box of chocolates. You never know what you're gonna get."
Forrest Gump

As I stood there dripping with algae, stinking to high heaven, I felt a little dazed. Jay Short and E. O. Wilson had disappeared with a small group of devotees trailing behind them to collect additional microbial samples from soils in Central Park's forests. Eric Mathur and Leif Christofferson were processing samples at the portable lab set up for the Central Park BioBlitz. The crowd that had been watching the dive team in the Harlem Meer was quickly dispersing. Perhaps they were disappointed that we had not found a body, or maybe they had been waiting around in case one of us drowned. You never know about New Yorkers.

Volunteers at the Dana Discovery Center provided us with a fat garden hose to spray down our scuba gear and shower the muck from our bodies as best we could. (The smell lingered for about two days.) I spread out my gear on the concrete, looked out at the park across the lake and began coming to terms with the fact that my job was nearly finished. Three years had passed since I went bug hunting in Costa Rica and stood in the surf as Mathur laid out the story of archaea and how its discovery had fomented a revolution in microbiology. Since then, we had collected samples from a mountaintop rainforest in Hawaii, from coral reefs in Puerto Rico, the vents at the bottom of Yellowstone Lake, and the hot springs of Kamchatka. We shared hotel rooms, meals and sunburns, and had gotten to see parts of each other's personalities that surface only under difficult field conditions. Mathur liked to tell people how I had been "embedded" with Diversa's expedition unit. But you shouldn't construe Mathur's remarks as my being "in bed" with the company. Ditto with any possible suspicion of my being in Craig Venter's pocket, despite my travels with him too.

Two weeks before the BioBlitz I sailed up the east coast with Venter aboard his *Sorcerer II* yacht. In January 1997, I sailed with him aboard the original *Sorcerer* five months after he announced the sequencing of *Methanococcus.* The fact is that Diversa and Venter's ventures are the leading enterprises in biotechnological developments from microbes. Diversa's collaborations with Eddy Rubin at the DOE's Joint Genome Institute (JGI) and with Jim Noble with the DOE's anti-proliferation program and the fact that Short's corporate ethics were respected by Yellowstone's John Varley provided considerable credibility for their work. Venter is controversial in this community but only in the sense that his style tends to rattle the cages of a relatively influential core group of academic and government scientists. Venter's accomplishments speak for themselves.

As I sat at the lake's edge waiting for my gear to dry, I pondered the many significant changes involving the Diversa crew and Venter that had occurred since returning from the trip to Kamchatka only 18 months earlier. None of the people I began this journey with in 2003 were still at Diversa. Venter had assumed a new central role in the story. And all of the same people had come together at a new horizon on this day of the launch of the E. O. Wilson Foundation.

Short resigned as CEO of Diversa in October 2005. He immediately began developing the E. O. Wilson Foundation and successfully launched it nine months later. If the foundation is successful, Short believes he will play a broader role than was possible at a private company in helping human society avert the rise of the second Archean Empire. One could argue that the most ancient life forms on the planet are poised to reclaim their territory as levels of carbon dioxide and methane rise in the atmosphere, as production of the ocean's phytoplankton continues to fall, and as our population and industries continue feeding sludge, plastics, metals, and all varieties of toxic wastes to the methanogens and other extremophiles. Should any natural disasters occur, such as the Yellowstone caldera turning into a supervolcano, archaea will gain an even greater advantage. Who knows what that would do to the global climate already undergoing a warming trend?

Short mentioned just a few days before the foundation's launch that he was becoming convinced that we had already passed a tipping point for global warming. If this is true, the big question is whether global warming will reverse itself in a few thousand years as it has done naturally in long cycles over time, or have human emissions pushed the climate too far? The latter wouldn't be good for life as we know it. The end stage of an irreversible warming trend is a carbon dioxide atmosphere, much like the one Earth possessed in its first 2.5- to 3-billion- year history. It's hard to say how plants would adapt, but everything with a set of lungs or a dependency on oxygen in other ways would fade into evolutionary history. The worst that would happen in the microbial world is that the cyanobacteria might have to give up the penthouse suite of microbial mats to the anaerobes, which would finally "breathe" in the open again.

Short's long-time vision has been to figure out how to get more people to realize the importance of microbes as the heart of our planet and their vital link in our continuing success as life's second domain. It worked on me. After spending so much time with Short, Mathur, Karl Stetter, and Venter, I no longer look at Earth the same way. When I see a hazardous waste site, I see a vibrant ecosystem for microbes that feed on poisons. When I see rust, I see colonies of iron-loving microbes. When I see dead grasses and a forest burned to oblivion by sulfuric acid and near boiling temperatures—as it is at Norris Geyser Basin—I see a breeding ground for strange new cooperatives of extremophiles. When Short looks at the same dead trees bleached white by hydrothermal fluids, he sees enzymes for bleaching paper in the wood pulp industry. I've actually developed quite a fondness for microbial life along with a new appreciation of their power.

It is easy to understand Short's drive to change public perception through the E. O. Wilson Foundation and to create new microbial-based economies. In my opinion, this is even more critical and unique. Short is using everything he learned over the past 15 years in the biotech industry to stimulate new mini-Diversas around the world. He intends to provide scientists in underdeveloped countries with the technical knowledge to conduct their own environmental sampling, microbial protein expression and gene sequencing, and teach them how to make money from their natural resources without destroying them. Wilson liked the idea so much he was willing to lend his name and legacy to Short's vision. To me that says a lot. In 20 years of covering science, I've never seen one of the top scientist/entrepreneurs in biotechnology team with an academic legend of Wilson's caliber.

"If dwindling wildlands are mined for genetic material rather than destroyed for a few more board feet of lumber and acreage of farmland, their economic yield will be vastly greater over time. . . . The wildlands are like a magic well: the more that is drawn from them in knowledge and benefits, the more there will be to draw," Wilson said.

The night before the BioBlitz, Short and I spoke together at a gala sponsored by the foundation at Central Park's famous Tavern on the Green. It was a bit surreal for both of us. We discussed with amazement about how quickly research on archaea and bacteria had advanced since the expedition to Costa Rica three years earlier and how dramatically the landscape in the biotech community had changed. Short was leaving the for-profit world behind as Venter was stepping into it—again.

"Diversa proved that the biodiversity of microbes can yield products and that the products can be competitive with other companies. We proved that you can partner with other countries and give royalties back. But it is challenging to go as far as you can in a company to make a real difference," Short said. "As a non-profit foundation we can try to engage other companies to do what we were doing at Diversa. If you can demonstrate how to make it profitable for them, they will do the right thing. Then you can revolutionize industry. A non-profit foundation has the ability to do these things. Companies need to be greater stewards of the planet but we all know a lot of them aren't. We have to educate people so that they do business with the companies that are aligned with conservation. If that is successful, the companies that are aligned with conservation will make more money than companies that are not. In industry it's about the money first and doing the right thing if it costs less."

Christoffersen had left Diversa to help Short with the new foundation. In fact, Christoffersen had been recruiting E. O. Wilson to serve as a scientific advisor to Diversa since 2003. Christoffersen had established a relationship with Wilson, which played a significant role in bringing together Short and Wilson in their new partnership. Mathur and his core scientific team from Diversa had left the company and joined forces with Venter at the Venter Institute (VI) and at Venter's new biotechnology company Synthetic Genomics Inc. (SGI). The last time Venter started a company, he sequenced the human genome. Venter had put together a team that included Diversa scientists as well as the top brass from the DOE's Microbial Genome Initiative, Aristedes Patrinos and Marvin Frazier. This was heady stuff.

Despite significant changes, everyone I started the journey with in 2003 was here at the Central Park BioBlitz in 2006, only in new roles. Mathur was still handling microbial samples, except now he was doing it officially for the VI, which was scientifically supporting the launch of the E. O. Wilson Foundation. The VI was sequencing the samples and providing analyses to Short, who was using the information to build a global microbial biodiversity database accessible to science teachers worldwide. The new database falls in line with the E. O. Wilson Foundation's primary mission of public education and outreach. Of course, if the Harlem Meer sediment has any good stuff in it, Mathur will help SGI find a way to genetically engineer it and make some money.

Looking back now, I saw the first hints of the changes to come just after we returned from Kamchatka. It was Monday morning, September 17, 2004. I was attending the Sixth International Genome Sequencing and Analysis Conference (GSAC) in Washington, D.C., sponsored by The Institute for Genomic Research (TIGR). E. O. Wilson gave the keynote address—he is obviously a sought-after lecturer. The morning session was packed with heavyweights. After Wilson, Venter spoke, and then Rubin followed by "Ari" Patrinos, assistant director of the DOE's Office of Biological and Environmental Research. Patrinos is one of the DOE's leading visionaries with respect to microorganisms. In 1994, he and Frazier, director of the Life Sciences Division at the DOE's Office of Biological and Environmental Research, started the Microbial Genome Initiative. Patrinos's scientific background is structural biology, global environmental change, nuclear medicine and health effects, and bioremediation of pollution. He also is a huge Rolling Stones fan and happens to be a celebrity in Greece. In 2005, a poll by the major newspaper in Athens voted him the fifth most important person of the year.

At the DOE, Patrinos and Frazier founded JGI in 1997 and also started the Genomes to Life Program in 2000. Patrinos and Frazier have made most of the key decisions since the early 1990s about which microbes the DOE would fund for whole genome sequencing. They gave Venter's scientific career a big boost when they approved grants in 1994 for the brand new TIGR to sequence microbial genomes. The *Methanococcus* project, which put archaea on the map in 1996, was essentially a collaboration between Venter, Woese, Patrinos, and Frazier.

With DOE funding, Patrinos and Frazier began fertilizing the phylogenetic tree planted in 1977 by Woese and nurtured along in the 1980s by Stetter, Norman Pace, and their academic offspring. In 1994, Patrinos and Frazier selected the first three microbes for sequencing with DOE funding: Stetter's *Pyrococcus furiosus*, because of its high temperature resistance; *Methanococcus jannaschii*, because it was sexy; and *Methanobacterium thermoautotrophicum* (actually a member of archaea), because of its potential to metabolize sewage sludge. Patrinos and Frazier were already thinking far ahead. They knew the best hope for cleaning up the messes created by the DOE and inherited from criminally negligent industries would be found in the metabolisms of microorganisms.

No one in the small world of microbial genomics could walk far without intersecting with the paths of Patrinos and Frazier. Similarly, no one could walk far without bumping into Short, Mathur, Venter, Stetter, and Pace—or least their work. It makes me think of pioneers of rock 'n roll and what a small world they started out in. The Beatles, Rolling Stones, and Beach Boys all listened to Elvis and to each other.

After Wilson's GSAC speech, Venter, who founded the annual GSAC in 1998, talked about his 2003 sailing expedition around the world collecting microbes from the oceans and sequencing their genes. Venter's expedition aboard *Sorcerer II* had found the diversity of the ocean to be far greater than imagined. It also created some tension at Diversa, which up to then had been unchallenged in the development of environmental sampling technologies and methods of turning that knowledge into products that make money. As soon as Venter established the non-profit VI and set sail, a number of people expected he would soon follow up with a new for-profit biotechnology company to capitalize on the microbial data being gathered. This was certainly apparent to Short because it would place the two of them in direct competition.

Venter's team collected and sequenced so many microbial genes during the global ocean expedition that the VI no longer had the technical and computational capability to sort through them and learn their function. Not enough scientists existed on the planet to determine the functions of all those sequences, and it would take some very bright minds to sort it out. Venter had not announced it yet, but he had just hired "Marv" Frazier away from the DOE to become Vice President of Research at the VI. This was a serious coup for the VI, and in accordance with Venter's style. He is an extremely loyal person and doesn't usually forget those who helped him on the way up. The VI presented an exciting opportunity for Frazier. He and Venter would soon set up a collaboration with the University of California, San Diego, and obtain a $24.5 million grant from the Gordon and Betty Moore Foundation. The project is called Community Cyberinfrastructure for Advanced Marine Microbial Ecology Research and Analysis (CAMERA). According to the collaborators, of the known DNA sequences currently sitting in databases, scientists understand the function of only about 40%. CAMERA will build the computational power to read and interpret the function of DNA sequences automatically. At the time, the project was focusing on the sequence data generated from microbial samples collected during Venter's ocean voyage. But once the technology is developed it can be adopted for any environment.

Venter's team published its first paper on the global sampling excursion in March 2004 in *Science.* He summarized the data again for the GSAC: more than 1.2 million previously unknown genes. Also identified were genes representing 782 new rhodopsin-like photoreceptors belonging to 1,800 new microorganisms, probably picoplankton involved in some vital but unknown way in the planet's uptake of carbon dioxide and output of oxygen.

"Microorganisms are responsible for most of the biogeochemical cycles that shape the environment of Earth and its oceans. Yet, these organisms are the least well understood on Earth, as the ability to study and understand the metabolic potential of microorganisms has been hampered by the inability to generate pure cultures. Recent studies have begun to explore environmental bacteria in a culture-independent manner by isolating DNA from environmental samples and transforming it into large insert clones," Venter said.

When Rubin's turn to present at GSAC came, he spoke about recent developments in microbial ecology and the emerging focus on the importance of understanding microbial metabolisms. Rubin's greatest contribution since taking the helm as acting director of JGI in spring 2002, and officially in January 2003, was expanding programs to sequence microbial communities, en masse, directly from the environment while continuing the work of sequencing individual microbial genomes that possess specific capabilities of interest to the DOE.

Rubin collaborated with a number of academic scientists, including Ed DeLong who had been doing microbial ecology and metagenomics in the oceans since the 1990s. Rubin's talk in 2004 was one of those milestone moments because the focus for years had been on sequencing genomes. Rubin was telling the crowd that microbial metabolism of communities of microorganisms was where he was steering the ship for the future at JGI. Adding metabolism to the picture pulls everything out from a two-dimensional perspective into full three-dimensional relief. At JGI, Rubin was applying his computational knowledge and insights to create a three-dimensional microbial universe. It's quite the thing to see how the genomes of different organisms interact with each other and react to each other's metabolites and to the geochemistry of the environment. The enormous capacity for gene expression and protein analysis that Diversa scientists were building had not gone unnoticed by Rubin. He had been making frequent trips to Diversa for two years to talk about their collaborations and develop the technology needed for the DOE's future initiatives.

JGI is worth paying attention to in the future. It is a well-organized and -funded scientific multiplex: One arm is involved with sequencing whole genomes of individual microbes—more than 100 so far. Another arm is focused on methods of analyzing sequence data—more, better, faster. Another focuses on eukaryotes while another examines the biology of genomes of all three domains. Still another one studies the evolutionary origins of genomes, and another still is devoted entirely to microbial ecology. JGI's many divisions and its collaborations with dozens of labs and several biotechnology companies have been extremely productive.

At the time of Rubin's talk at GSAC he was completing a proof-of-concept study to show how gene expression and sequencing could be used to determine the metabolic signatures of different environments. JGI and its collaborators, including Diversa, were working on a landmark paper that would be published in *Science*. The study compared the metabolic signatures of microbes from three different environments—a contaminated mine that had become extremely acidic; ordinary farm soil; and several ocean bottom environments where dead whales had found their final resting places and were being metabolized by microbes. The research clearly demonstrated that it is possible to create a metabolic profile of any environment using only random fragments of raw microbial DNA. The metabolic profiles, which Rubin called functional fingerprints, could be used to provide highly accurate assessments of the vitality of environments such as forests, lakes, rivers, farms, and specific regions of the world's oceans. The metabolic profile of an area reflects the presence and levels of nutrients, pollutants, or other environmental features of interest. It can reveal environments currently under stress as well as reflect progress in environments undergoing remediation.

Rubin said the "functional fingerprinting" methods were comparable in significance to the development of expressed sequence tags, which Venter pioneered in 1991. JGI called their signatures environmental genome tags (EGTs). The EGTs, which were developed in part through Mathur at Diversa, reflect the level of diversity of microbial populations as well as the different types of gene families present. Rubin told the audience the EGT information could be pieced together to form the bigger metabolic picture of an entire environment. EGTs are quickly becoming an important tool for microbial ecology. They have the potential in the near future to diagnose sick environments or monitor healthy environments located near industries, businesses, or government facilities that have the potential to pollute soil and groundwater. Personally, I would want one of these devices. I recently watched a community of expensive town homes being built on a former railroad yard where hundreds of old wooden rails and telephone poles had been rotting for more than a decade. If I were buying a home there, and before I ever let my children out to play, I would want a metabolic profile made of the soil to determine the presence of dioxins and other contaminants. I bet this type of technology when it becomes commercially viable will be quite appealing to the public.

"The EGT fingerprints may be able to offer fundamental insights into the factors impacting on various environments," Rubin said. "With EGTs we don't actually need a complete genome's worth of data to understand the functions required of the organisms living in a particular setting. Rather, the genes present and their abundances in the EGT data reflect the demands of the setting and, accordingly, can tell us about what's happening in an environment without knowing the identities of the microbes living there."

EGT fingerprints capture a DNA profile of a particular niche and reflect the presence and levels of nutrients, and pollutants, as well as features like light and temperature. For example, genes involved in breaking down plant material are over-represented in soil and absent in the seawater, while in seawater, genes involved in the passage of sodium, a major chemical component of saltwater, are particularly abundant. As light is a major energy source for microbes living in surface water, there was an abundance of genes involved in photosynthesis in samples collected from shallow water. These differences in the abundances of genes involved in particular functions provide DNA clues to features of the environments from where the samples were taken. Importantly, the DNA clues were easy to find despite the vast numbers of different individual microbial species within the samples.

Rubin argued that the genes and their relative abundance as gleaned from EGTs reflect the intricate physical and biochemical details of a given environment without having to know the actual identities of the resident microbes. This finding is crucial for studies of microbes in the wild since the sheer number of different organisms present in nearly all environments makes it a daunting task to sequence the multitude of organisms one at a time. With the EGT approach, an abbreviated sequencing effort enables scientists to piece together the information and form a useful metabolic picture of an entire complex environment.

At the time of his talk Rubin was working on a pet project of his own, obtaining tiny bits of Neanderthal DNA from fossilized bone and piecing the gene fragments together into a more complete picture of the Neanderthal genome. The work is interesting on its own, but it has implications for ancient microbial DNA. Rubin and collaborators at the Max Planck Institute in Leipzig, Germany, are now comparing Neanderthal genes to modern human genes. Their goal is to eventually obtain enough Neanderthal DNA to map most of its genome. Rubin's team tested their methods of sequencing ancient DNA first on the fossilized bone of an extinct Pleistocene cave bear species. They published a study on the cave bear DNA in June 2005. Ancient DNA poses special challenges because DNA starts breaking down at the moment of death. Microbes go straight to work recycling the organic nutrients in the newly deceased body. Currently, the theoretical limit for sequencing ancient DNA is about 100,000 years for samples preserved in the same conditions in which the cave bear specimens were found—relatively dry, high altitude, with moderate temperatures—or if frozen, perhaps longer, according to Rubin.

Patrinos was next up at the podium. I was settling in for his usual PowerPoint presentation on how the DOE's work would someday save the world, when remarkably I heard him say:

"I no longer feel that I have to hold my nose when I tell people that research on microbial genes and microbial systems will help us answer questions of tremendous significance to the future of society. We have been making advances over the past couple of years that make me really believe it," Patrinos said from the podium.

I leaned over and whispered to Christoffersen, who was sitting next me in the audience, "Did he really say that? 'I no longer have to hold my nose'?"

"Yeah," Christoffersen said, surprised. "I think he's saying it isn't bullshit anymore."

When Patrinos shocked me to attention I looked up and saw on the giant screen behind him a slide of the DOE's Hanford Site in the state of Washington. The statistics are horrifying. Of 149 tanks storing toxic waste, 65 are leaking. This is an archaean paradise: 35 million gallons of waste, 190,000 tons of chemicals, 13 million curies of radioactivity, 29 million cubic meters of contaminated soil, 39 million cubic meters of uranium mill tailings, and 4.7 million cubic meters of contaminated groundwater. And this is just one of hundreds of lethal messes the federal government is responsible for creating and eventually cleaning up around the United States.

"This material, this witch's brew, is moving. You can't just put a fence around it. It is migrating into aquifers and rivers," Patrinos said candidly.

Patrinos explained that for more than 50 years "the United States created a vast network of facilities for research and development, manufacture, and testing of nuclear weapons and materials. The result is that more than 7,000 sites at over 100 facilities across the nation have subsurface contamination, more than half of which contains metals or radionuclides and most including chlorinated hydrocarbons. DOE is responsible for remediating two trillion gallons of contaminated groundwater and 70 million cubic meters of soil and subsurface sediment. The groundwater volume is equal to about four times the U.S.

daily water consumption, and the sediment volume would fill 17 professional sports stadiums. The DOE estimates that, using current technology, cleanup will take 70 years at a cost of $300 billion."

Next he showed a graphic of global carbon emissions and the increases of carbon dioxide levels in our atmosphere over the past 50 years. This also was disturbing.

"The DOE also has some responsibility for global climate change. There is an increasing suspicion that emissions of greenhouse gases are contributing to the effect known as global warming," he said not so candidly. Even if Earth is experiencing a natural global warming cycle, most scientists are pretty sure man-made emissions are accelerating it.

Patrinos displayed an idealized graphic of microbes being used to convert corn into energy, microbes being used to generate hydrogen fuel, and microbes metabolizing hazardous waste with by-products that are less toxic and more easily cleaned up. None of this is being done yet on a large scale because it is much more difficult to do than the graphic suggests. However, it is possible that by "harnessing" microbial genes, biotechnology could save the U.S. economy tens of billions of dollars annually, clean up the environment by 2030, stabilize and cut emissions of carbon dioxide into the atmosphere by 2040, and produce hydrogen energy on a full scale with engineered microbes by 2050. It is the harnessing part that poses the greatest challenge. What Diversa is doing with microbes to create specific enzymes with specific properties for specific purposes is relatively easy by comparison.

Next Patrinos outlined how microbes constitute most of the living matter on Earth and have been evolving for nearly 4 billion years (at least). He brought up how archaea and bacteria have been found in virtually every environment, thriving in extremes of heat, cold, radiation, pressure, salt, acidity, and darkness. He explained how no other forms of life can obtain nutrients and energy directly from inorganic matter and how important it is for scientists to learn as much as possible about how microbes perform these chemistries. This is all absolutely true. He emphasized the importance of studying the biodiversity of microorganisms and the range of their environmental adaptations. Again, spot on if we are ever to actually harness microbes for our benefit.

"Microbes long ago solved many problems for which scientists are still actively seeking solutions," Patrinos said. "Researchers have only scratched the surface of microbial biodiversity. Knowledge about the enormous range of microbial capacities has broad and far-reaching implications for environmental, energy, health, and industrial applications."

Patrinos displayed a few more graphics to show how microbes can be used for bioremediation. Then he dropped another great line.

"A few years ago when I had to present this information I felt that it was science fiction. Now I believe it. All it takes is funding," Patrinos said. "If we were to find life on Mars today, Congress would write us a blank check to fully investigate the microbial world—the invisible world here on Earth. Yes. That would do it."

It is true. A focused and funded commitment to microbes similar to the Manhattan Project or the *Apollo* Moon Mission would revolutionize the world we live in today. The surface of microbial biodiversity has only been scratched, but scratched deep enough to see tremendous power in the genes and metabolisms of bacteria and archaea.

Absent the blank check, what's likely to be coming up in the near future? I believe Diversa and SGI and the few companies like them hold great promise in the biotech industry to make a difference quickly. Drug discovery faces so many regulatory and scientific hurdles that progress with biomedical applications will remain slow. But there's no doubt that the concept of using microbes to produce enzymes for industry works and faces fewer regulatory hurdles. The impact will be limited until more companies get into the business of microbes. This is where the E. O. Wilson Foundation can have a significant influence if it is able to raise enough money to stay afloat and achieve its goal of reaching the public. The real power to change economies lies with the consumer. Just as people are beginning to demand hybrid cars and organic foods, the public could demand that products made by the chemical and agriculture industries be made, when possible, with microbial enzymes that are safer and more eco-friendly.

As a single company, perhaps SGI holds the greatest power. SGI has too many good scientists not to succeed in some way. Venter officially launched SGI on June 29, 2005. On February 2, 2006, Patrinos signed on as the company's president. The combination of scientific expertise and power at the VI and SGI is simply unprecedented.

I can attest to the fact that SGI was already a twinkle in Venter's eye as early as 1995. That year, Venter and Nobel Laureate Hamilton Smith, then at Johns Hopkins University, announced completion of the first microbial genome *Haemophilus influenzae*, which had taken them a year to figure out using TIGR's own money. But that work proved the capability of shotgun sequencing of whole microorganisms. Patrinos and Frazier supported TIGR's work on the second microbial genome, *Mycoplasma genitalium*, which was accomplished in only three months. *Mycoplasma*'s genome turned out to be a surprisingly tiny thing, which gave Venter and Ham Smith a few ideas that set the course of their future. Despite sequencing the human genome, Venter's first scientific love was and still is microbes.

After discovering what a small genome *Mycoplasma* possessed, Venter and Smith began wondering what might be the minimum amount of DNA essential to life. A minimal genome would reveal some of nature's top secrets about the origin or origins of life and it might reveal how to build an organism. TIGR reported the sequencing of *Mycoplasma genitalium* on October 20, 1995, in *Science*. It is a urinary tract bacterium that possesses only 517 genes.

Venter envisioned what might be the minimal genome necessary to create an artificial organism for some time. He said publicly shortly after publishing the paper on *Mycoplasma* that he didn't think it would be all that difficult to build an artificial organism. The trick would be splicing the right genes and getting them to recognize each other as a string of code. Smith had won his Nobel Prize for understanding the basics of how DNA does this naturally. It you wanted to look back and find the genesis of SGI, it was with the sequencing of *Mycoplasma genitalium*, made possible by Patrinos's and Frazier's support and their own curiosity about the criteria for a minimal genome. For a microorganism all you need to do is stick the genes of the artificial chromosome into a vector such as a virus and express the stitched genes in *E. coli*. Voila. It's not life in the self-replicating sense, but it's a start. To be life it must self-replicate and it must be able to mutate. I wondered in what ways an artificially created genome would mutate?

"Frankencell," Venter called it. Venter and Smith started experimenting right away on eliminating genes from *Mycoplasma* to learn how many genes and which ones could be knocked out and still leave them with a functioning organism. Venter reported in 1999 that possibly 250 to 300 genes could define the minimal genome for life.

The next milestone reached toward creating Frankencell occurred in the summer of 2003. Patrinos and Frazier funded a grant at Venter's non-profit institute to construct an artificial genome. Venter's group decided on creating an artificial *viral* chromosome to start with. Viruses are simple and easy to express in bacteria. Construction of an artificial polio genome was reported in 2002 by another group, making it technically the first, but it took those scientists three years to do it. Venter's group spliced their synthetic fragments of DNA together in two weeks using new methods and more powerful tools. Technically, though, the work began in Venter's head in 1995.

Elizabeth Pennesi covered the artificial chromosome story for *Science* in November 2003. She wrote, "they started with short pieces of DNA, pieced them together by matching up overlapping ends, and eventually generated a complete 5,400–base-pair phage genome. . . . And Venter is convinced that he can build genomes 300,000 bases or longer. But even with these improvements, skeptics and supporters aren't sure how well the procedure will work for organisms with larger genomes. 'Going from a phage to a microbial genome to having a microbe that's synthetic is a very major step,' says Patrinos. But he thinks it's worth betting on."

The success of the artificial chromosome project gave Venter enough proof of concept to begin planning SGI. It was exciting to watch the idea unfold. It was just about all he wanted to think about for the next two years. It looks like Patrinos staked his career on SGI being a good bet.

SGI has been extremely coy about its specific plans. Not even Mathur will spill the secrets to me. Friendship only goes so far. But secrecy is typical behavior for any new biotech company. This is a viciously competitive arena full of corporate espionage and intrigue. What SGI *is* saying in its corporate boiler plate is that the company plans to "design, synthesize and assemble specifically engineered cell level bio-factories. The ability to make extensive changes to the DNA of a chromosome, assemble it, and insert it into an organism is in its infancy, and the capability to assemble chromosome length strands of DNA will be key to the success of the company."

If you recall, Diversa was designing new enzymes by automating permutations of combinations of amino acid sequences. When they had something of interest, they would synthesize the new genes and pathways and express it in a bacterial cell, which worked as a biofactory. SGI is essentially taking this to the level of nucleotide sequences to craft whole chromosomes for specified industry goals. The synthetic chromosome or set of designer genes are expressed in a similar cellular platform. According to SGI, "Using the genome as a bio-factory, a custom designed, modular 'cassette' system will be developed so that the organism executes specific molecular functions. Synthetically produced organisms with reduced or reoriented metabolic needs will enable new, powerful, and more direct methods of bio-engineered industrial production. After designing and producing a synthetic chromosome, the team plans to develop a proof of concept in either of two bio-energy applications—hydrogen or ethanol. We believe that the synthetic chromosome, and eventually a synthetic cell, will become an integral tool for the energy industry." Frankencell!

Recall Mathur's fascination with termite hindgut microbes. This work is beginning to bear significant fruit. JGI announced in the summer of 2006 that it completed the sequencing of termite hindgut species. A number of new projects involving the breakdown of cellulose for biofuels and termites stole the headlines. It is exciting to see bugs making the news. The termite hindgut sequencing was the direct result of Mathur's trip to Costa Rica in April 2003. Rubin had liked Mathur's termite idea from the start. Termite hindgut microbes fit nicely with the DOE's mission to develop new bioenergy fuels. At the same time that Rubin selected the termite hindgut microbes for sequencing, he approved a project to sequence the genome of a species of Poplar tree. That project was completed in September 2006. The black cottonwood, *Populus trichocarpa*, is a potential "feedstock" for a new generation of biofuels such as cellulosic ethanol, Rubin said. The tree can grow about 12 feet per year. Forests of these trees could be grown in a few short years and continually harvested. Combine that with enzymes made by microbes in termite hindguts and you've got a new, natural, and renewable biofuel industry.

This may sound a little boring at first, but here's a tip: prairie grasses. They are one of the most promising new sources of biomass for making biofuels. While prairie grasses do not yield quite as much biofuel—gallon for gallon and acre for acre—they are easier to grow and much less expensive to cultivate than corn. Recent research suggests that prairie grass—dollar for dollar spent on cultivating and processing—will yield five times more fuel than corn.

JGI announced on May 23, 2006, the completion of the sequencing of its 100th microbial genome. The pace of microbial sequencing, and metabolomics, has jumped dramatically since Venter's sequencing of *Hemophillus influenza* in 1995. By spring of 2007, JGI expects to have at least 150 microbial genomes finished.

"The power of DOE JGI sequencing microbes, and other organisms, is that it gives us the complete genomic 'parts list' of those organisms," Dr. Raymond L. Orbach, Director of the DOE Office of Science, said at the announcement of the 100th genome. "With this list in hand, we can explore how microbes use these parts to build and run their key functions, many of critical importance to DOE because they can break down plant materials to produce such useful sources of energy as ethanol and hydrogen, and clean up toxic waste sites. We know that microbes can perform these and a multitude of other amazing tasks and with the proper technology we can harness these capabilities."

Before joining Venter, Patrinos and Frazier had developed with Rubin a long list of microbial species for whole genome sequencing. These microbes have been studied independently by a number of teams at different universities and government laboratories over the years for their potential to metabolize key pollutants and generate energy. With whole genomes in hand, scientists can understand more precisely how the microbes do what they do. Then labs can engineer the genomes to make bugs more effective or use the genes to create enzymes or products to catalyze reactions *in situ* in polluted environments, or in the big storage tanks that are leaking the witch's brew.

Bioremediation has been a long-time goal of the DOE. It will be the hardest to accomplish. Quite recently, JGI released a list of new organisms that it believes will be useful to clean up specific types of pollutants.[1] All of these are bacteria. One of the reasons they selected mostly bacteria for the next round of sequencing is that the organisms are easier to work with. Archaea are still more difficult to culture and maintain in culture because of their cellular differences. Among some of the key microbes selected for whole genome sequencing that could revolutionize our future:

- *Burkholderia* species have been studied by a number of teams, which have shown that these protozoa are efficient degraders of pollutants in water and soil, and naturally assist in fixing atmospheric nitrogen to the roots of plants. They defend their host plants against pathogens. These traits make *Burkholderia* valuable players in a healthy, carbon dioxide–cycling ecosystem. The species are abundant in soil around the roots of plants, in streams, and in lakes and they have developed numerous partnerships with plants, animals, and even amoebas. According to the DOE, *Burkholderia* are major nitrogen-fixing partners of plants in the two largest CO_2-sequestering ecosystems in the Western Hemisphere, the Amazon rainforest and the Cerrado Savannah of Brazil. Jim Tiedje at Michigan State University is leading studies to use *Burkholderia* to break down polychlorinated biphenyls.

• *Chromohalobacter salexigens* is a salt-loving "proteobacterium" that displays resistance to heavy metals and has an ability to degrade aromatic hydrocarbons. It is related to *Pseudomonas* and *E. coli.* Laszlo Csonka of Purdue University; Brad Goodner of Hiram College; and Aharon Oren of the Hebrew University of Jerusalem are leading studies of its ability to clean up toxic organic pollutants.

• *Deinococcus geothermalis* is one bad son of a bug. It is extremely resistant to radiation and a real troublemaker. It is a bacterium that creates pink biofilms that resemble a pearl necklace. It attaches itself to steel and eats holes in it. It is a considerable nuisance to the paper industry and nuclear power plants, according to Christina Saarimaa of the Bioprocess Engineering Laboratory at the University of Oulu, Finland. It withstands enormous amounts of radiation, dryness, and heat. Because of its natural talents it is being engineered to metabolize mixed types of radioactive waste at temperatures up to 131 degrees F. Studies are being led by Michael Daly of the Uniformed Services University of the Health Sciences; James K. Fredrickson of the Pacific Northwest National Laboratory; and Kira S. Makarova of the National Institutes of Health.

• *Desulfovibrio desulfuricans* has an appetite for sulfur, uranium, and toxic metals. On its own, it corrodes iron in pipeline, and injects hydrogen sulfide gas into petroleum, which makes gas and oil smell like flatulence among other problems. In 1992, Derek Lovley of the University of Massachusetts conducted studies on this microbe and suggested that it "might be a useful organism for recovering uranium from contaminated waters and waste streams." We've known about this for a decade and a half. Judy D. Wall of the University of Missouri is leading studies of this one.

• *Geobacter metallireducens* is an underground "metal head" discovered in 1987 in sediments of the Potomac River less than half a mile from my apartment in Alexandria, Virginia. It was the first microbe found to oxidize organic compounds to carbon dioxide using iron oxides, essentially rust, as a source of energy. If it can't find metal, it grows flagella and tracks it down with sensors in its cell. It is anaerobic, loves the dark, and is an important natural player in the carbon and nutrient cycles of aquatic environments. Lovley is leading studies to develop its talents to clean up organic and metal contaminants in groundwater. I wonder if this species will be found among the many involved in rusticles. According to Lovley, "One of the most exciting discoveries in the past few years in microbial fuel cell research was the development of a microbial fuel cell that can harvest electricity from the organic matter in aquatic sediments. These systems are now known as Benthic Unattended Generators or BUGs." *Geobacter* has the ability to directly transfer electrons to the surface of electrodes when placed in seawater between dissimilar metals. This is exactly how microbes on the *Titanic* are getting power, and apparently generating more of it. BUGs was developed by Clare Reimers of the College of Oceanic and Atmospheric Sciences at Oregon State University and Leonard Tender of the U.S. Naval Research Lab. BUGs are being designed for powering electronic devices in remote locations, such as the bottom of the ocean, where it would be expensive and technically difficult to routinely exchange traditional batteries, Lovely said. He said similar designs could potentially power electronic devices in remote terrestrial locations and could even eventually be modified to harvest electricity from compost piles, septic tanks, and waste lagoons. Imagine the potential for rural and underdeveloped areas.

- *Rhodobacter sphaeroides* is one of those microorganisms that simply must be sequenced because it is so unusual. Samuel Kaplan of the University of Texas Health Sciences Center at Houston, has been publishing studies on the microbe at least since 1972! According to Kaplan, *Rhodobacter sphaeroides* is among the most metabolically diverse organisms known, capable of a wide variety of growth conditions. "For example, *R. sphaeroides* possesses an extensive range of energy acquiring mechanisms including photosynthesis, lithotrophy, aerobic and anaerobic respiration." It can also fix molecular nitrogen and synthesize important pigments, chlorophylls, heme (used in energy transfers), and vitamin B_{12}. *Rhodobacter* also is a communicator; a "social" bacteria. It is the first free-living bacterium known to utilize the regulatory systems associated with quorum sensing—the ability of microbes to communicate with each other to coordinate group behaviors. It also has an almost unbelievable repertoire of ways to manipulate coding in its genes and transfer whole genes around in chromosomes. These traits make the organism a potentially useful tool for genetic engineering as well as for explaining a lot about the evolution of genes. Kaplan said one variety has been shown to detoxify a number of metal oxides, making it potentially useful for bioremediation. If one could combine quorum sensing with bioremediation, the result would be an army of microbes that could respond to human-generated signals. Finally, the bacterium has many traits similar to the mitochondria of our own cells. Perhaps it is ancestral and played a significant role in the evolution of eukaryotes. The DOE believes it has utilitarian potential also for light-driven, renewable-energy production and can detoxify metal oxides.

- *Shewanella* species. What can you say about these microbes?[2] I mean, they are superstars. Kenneth Nealson of the University of Southern California pioneered work on this organism. Nealson and Venter collaborated on the first *Shewanella* genome sequence, sparking the interest of the DOE. Jim Fredrickson at the Pacific Northwest National Laboratory (PNNL) is leading studies on the power of *Shewanella* to degrade metals including uranium, technetium, and chromium. *Shewanella* is a vital player in the carbon cycling of anaerobic environments. It converts soluble metals and compounds, like uranium, into insoluble forms, which gives them the ability to metabolize the nuclear waste that is contaminating groundwater. *Shewanella* takes the bad uranium into its cell, converts it into energy to make its cellular components, and then respires bad uranium in a precipitate that falls out of the water column and into sediment. *Shewanella* can live aerobically or anaerobically and can grow naturally almost anywhere. They do not cause disease in humans, animals, or other organisms. *Shewanella oneidensis* strain MR-1 is a leading candidate for cleaning up the DOE's contaminated nuclear weapons manufacturing sites, according to Yuri Gorby, also at PNNL. "We would like to find out how *Shewanella* works so it can be used to help remediate contaminated DOE sites. We also believe that understanding *Shewanella* may help explain how early life on Earth developed and functioned." *Shewanella* is an iron-lover. Maybe it too is on the *Titanic.* Gorby suggested *Shewanella* is a good model for life that might be residing on iron-rich Mars.

Meanwhile, scientists at many other labs are busy at work using multiple sources of funding. TIGR recently deciphered the genome sequence of a microbe that can be used to clean up pollution caused by chlorinated solvents—a major category of groundwater contaminants often left behind as by-products of dry cleaning or industrial solvent production. The study of the DNA sequence of *Dehalococcoides ethenogenes* found evidence that the soil bacterium may have developed the metabolic capability to consume chlorinated solvents fairly recently—possibly by acquiring genes as an adaptation related to the increasing prevalence of the pollutants. This is amazing! Natural selection on speed.

"The genome sequence contributes greatly to the understanding of what makes this microbe tick and why its metabolic diet is so unusual," according to TIGR scientist Rekha Seshadri.

The microbe was discovered initially by Cornell University scientists at a sewage treatment plant in Ithaca, New York. It is the only microbe that is known to convert the pervasive groundwater pollutants tetrachloroethylene (PCE) and trichloroethylene (TCE) into a nontoxic by-product. Currently, environmental consulting companies are attempting to use the microbe to clean up at least 17 sites contaminated by PCE or TCE in 10 states, including Texas, Delaware, and New Jersey.

Scientists at the federal Idaho National Engineering and Environmental Laboratory in Idaho Falls discovered by accident an archaeum in a hot spring at Yellowstone National Park that may clean up industrial wastewater generated by paper and clothing manufacturers before releasing the wastewater back into rivers and streams. In experiments, a protein contained in the microbe was extracted and added directly to the wastewater. The protein breaks down hydrogen peroxide, used to bleach clothes and paper. The Yellowstone protein worked 80,000 times longer than chemicals currently used to clean up hydrogen peroxide. The scientists accidentally stumbled across the microbe in samples they collected from a Yellowstone hot spring on an expedition to find microbes that would help in processing sugar beets, according to a report by Genome Network News. Instead of sugar beet solutions, they found a microbe that breaks down hydrogen peroxide at high temperatures and high pH.

Based on what I've learned on this wild ride, research on microbial consortiums—as opposed to individual microbes—offers the greatest promise for the future. The most interesting work of this nature is taking place in the nastier sludges and toxic waste environments of our little planet, where bacteria and archaea are commonly found working hand in glove to break down various organic and inorganic compounds. A term that keeps popping up in the scientific literature to describe the working relationships of these microbes is "syntrophic."

The Darwin Center for Biogeology in The Netherlands recently released a striking paper about a previously unknown microbial consortium that coupled anaerobic methane oxidation to denitrification in soil. The Darwin Center described the study as revolutionary. It found a bacterial species and an anaerobic methanogenic archaeum oxidizing methane in a manner never seen before. Normally the oxidation of methane requires oxygen or a sulfate. Instead, the consortium was using nitrite or nitrate. The team has since found examples of this partnership in samples of soil worldwide. Aside from being something completely different, the findings have significance for modeling climate change involving the greenhouse gas methane, according to Mike Jetten and Marc Strous of the Radboud University, Nijmegen, and the Darwin Center for Biogeology.

In a feature interview provided by the Darwin Center, the team concluded:

> The bacterium and the archaeon each play their own part in oxidizing methane. They're the missing link in the natural methane cycle, which plays an important role in our climate, explained Strous. Methane is a major greenhouse gas which is released by the anaerobic decomposition of dead plants at the bottom of a ditch. The more nitrate that is emitted; the more methane ends up in the atmosphere. Our discovery of this "methane glutton" shows that the cycle is in fact completely different from what we previously thought. Once we understand exactly how the cycle works, it will be a lot easier for us to predict how the climate will be affected and how we might intervene, if necessary.

Jetten said, "The discovery shows that we have still got a lot to learn and that bacteria are capable of all sorts of conversions we used to think were impossible. It means that the carbon and nitrogen cycles are in fact related in a completely different way. This may have consequences for our current climate models. The knowledge and experience we gained from our research into the anammox bacterium helped us to identify this microbial consortium. In the research we are doing within the Darwin Center we will be investigating the commonness of this process and the role it plays in the nutrient cycle."

Thirty years of research on archaea paved the way for modern microbiology to dramatically change the way we live in the world. Syntropy seems to be the hallmark of microbial behavior. As microbiology moves forward in its exploration of consortiums, it is certain to find many more unusual partnerships. Imagine what a concentrated research effort into the cooperative relationships of microorganisms might accomplish in the near future. Let's make plastic products with natural toxins instead of synthetic ones and then feed the pollution to other microorganisms instead of our children. Let's demand that our industries begin adopting the new and potentially greener chemistries that microbes have to offer us. We could be generating more electricity with giant BUGs instead of coal. We could do so many things with these champions of syntropy. But, one thing is fairly certain. Congress is unlikely to fund microbial projects at a level to make real history or even meet the conservative timelines Patrinos outlined earlier unless someone does find life on Mars, or we run out of oil, or consumers finally do demand new products and fuels.

Where we will find the greatest technical challenge is in efforts to use living microbes for bioremediation. So far, efforts to harness microbial genes and engineer microbes to clean up hazardous waste have met with limited success. These projects have not been flops. They are just much more difficult to pull off than scientists had initially imagined. The main problem is the organisms that are being introduced into target sites die off pretty quickly. This suggests that the engineered microbes are unable to integrate into the microbial community that is already present in the environment and they are unable to obtain enough nutrients to establish colonies and survive. Microbes function as communities and are often, if not usually, interdependent on each other's metabolites for survival. Until scientists have worked out these complex three-dimensional relationships, efforts at introducing engineered microbes will continue to meet with limited success.

One alternative suggested by Norman Pace is changing the chemistry of the environment to favor specific microbes that are already present. Early research is under way to see whether this might be a more viable approach in the short term for using microbes to clean up specific waste sites.

The challenges highlight the importance of Rubin's focus on metagenomics of communities and their metabolic signatures. It has only been in the past few years that scientists have begun to understand how microbes interact in the wild. In September 2006, JGI announced completion of a project that highlights the importance of understanding the cooperative behavior among different organisms.

The target of the project was the "gutless" worm, which has no mouth or anus. All of its energy requirements are performed by microbes that literally crawl into its skin and do all the metabolic work for it. In return the worm transports the microbes to sources of nutrients. This may represent a very ancient form of cooperation between early multicellular life and microbes. The worm, *Olavius algarvensis*, was found in marine sediments in the Mediterranean Sea. The microbes absorb carbon and synthesize all required amino acids and vitamins and deliver them to the cells of the worm. Waste products including ammonium and urea, generated by the worm's cells after using the energy, are taken back up by the microbes, which they use as a source of energy to obtain nitrogen.

"It's an excellent example of outsourcing energy and waste management, where this worm and the microbes living under its skin are enjoying a mutually beneficial relationship," said Rubin. "The microbes, floating around in the sea, strike up a bargain with the worm—in exchange for housing, the microbes take care of energy production and handling the waste."

DOE characterized the work as "the first instance of such a symbiotic relationship being analyzed by using a metagenomic shotgun sequencing approach, heralding a renaissance in symbiosis research."

Man, these guys have to get their hands on some rusticles! When the first rusticles were brought up from the *Titanic*, Cullimore and Pellegrino discovered that the system of arteries that channel nutrients throughout the *thing* were filled with hundreds of worms. They spilled out onto the deck with a foul odor. At the time Cullimore thought the worms might be providing a service, like Roto Rooter, to keep the water channels of the rusticles open. Could this have been the same species of gutless worm, incorporating rusticle microbes into its body? It would have been a feast for all plus an effective mechanical service for the *thing*.

"It's not unlike a car with a hybrid motor that can run on both electricity and gas depending on the situation," Rubin said. "In certain places the worm is powered by specific bacteria that can exploit the chemical energy abundant at a specific location, while in other strata, where a different chemical energy source is abundant, the worm switches its energy production to resident bacteria that can exploit that available energy source. We could learn something from this relationship. We've been dependent on fossil fuels. In the future we need to adapt like this worm has and use a variety of different energy sources to ensure our needs can be met in a changing world."

Symbiotic relationships and microbial consortiums appear to be an especially common feature of stressed environments. Consortiums arise in areas where nutrients are less available and especially in environments where pollutants have been introduced. That they arise under stressful conditions suggests the powerful and innate role of cooperative behavior in the very early development of first life. Cooperation through horizontal gene transfer in primitive nondescript cells could have led to the first organisms and eventually to the three known domains of life. Many scientists have long suggested that cooperative behavior among microorganisms led to the development of eukaryotic cells, which contain bacteria-like mitochondria. The mitochondria make energy for the eukaryotes much like the microbes in the skin of the gutless worm make its cells energy. Eventually, the same cooperative drive led to even greater complexity in life forms, such as microbial mats, spires, biothems, and the wondrous near-century-old consortium on the *Titanic.*

If I were to begin right away on a sequel to this book, I would focus entirely on the cooperative behavior of microbes, and on the factors that give rise to microbial consortiums of any type. I am especially curious about where the innate drive to work together to maximize efficiency at the most fundamental metabolic level originates. You can bet that the central focus of microbial research in the next decade will be on microbial communities and what makes them tick. If I had Rubin's power, I would find out immediately what is going on in terms of metagenomics with the *thing.* And if I had Venter's power and funding, I would conduct my next sampling expedition outside the International Space Station. Apparently, Earth is continually bombarded with particles of ice. Imagine designing a very fine seine like those small enough to capture viruses in the ocean and stretching it out like a big net in Earth's orbit. It might capture some interesting material that scientists aboard the space station could investigate for evidence of primordial matter and perhaps spores of extraterrestrial microorganisms carried through space on cosmic winds. If life exists beyond Earth, then it must be finding all types of creative ways to survive and flourish. Rubin's gutless worm and the rusticles are mere hints at the creative potential of life. At the heart of all of it is that unfaltering signal that shouts, "carpe diem!"

Modern microbiology is developing the tools to finally explore the microbial universe on Earth, teach us how to harness it, and perhaps discover where life came from. It is mind-boggling to consider how much the ideas that scientists have held about the origins of life and the tree of life have changed since Woese began asking, "What is it?" At least we can begin to admit that much of what we thought we knew about life 30 years ago was wrong. Perhaps the closer we peer into the microbial universe the more our minds will open to the bigger one beyond Earth. Whatever happens in the future, one thing is certain. The resistance to entropy will remain alive and well on Earth with us or without us. We might easily perish after having spent just a few tens of thousands of years on Earth as a species. But life will continue. The movement may be at work on many other worlds as well. Metabolism and cooperation seem too grand of things to be limited to a single tiny world.

This is only a hunch, but I think eventually we will find the secrets to life's origin buried deep in the hearts of those big incubators we call stars.

NOTES

CHAPTER 1

1. Paul W. Lepp, Mary M. Brinig, Cleber C. Ouverney, Katherine Palm, Gary C. Armitage, and David A. Relman, "Methanogenic Archaea and Human Periodontal Disease," *Proceedings of the National Academy of Sciences* vol. 101, no.16 (April 20, 2004):6176–6181.

Some people don't get excited about microbes unless they cause disease. If archaea somehow learned to tap the iron in our blood as a source of energy, these underappreciated microorganisms would get quick attention. For now, consider that one of the most adaptive, extreme forms of life is colonizing the dark, anoxic recesses of your gums. Perhaps only a single random gene mutation is needed for it to become the vampire of microbes and begin feeding on our blood.

2. Kenneth Todar, *Todar's Online Textbook of Bacteriology*. Todar is a professor in the Department of Bacteriology at the University of Wisconsin-Madison. He provided me with tremendously valuable materials on metabolism. His time spent creating a freely accessible online

textbook is an enormous public service. Todar has taught microbiology courses at the University of Wisconsin-Madison for 31 years. His main teaching interests are in general microbiology, bacterial diversity, microbial ecology, and pathogenic bacteriology. The lectures and readings from his courses from which his textbook derives can be found at http://www.bact.wisc.edu/themicrobialworld/homepage.html. The online textbook is at http://www.textbookofbacteriology.net/.

3. John L. Howland, *The Surprising Archaea: Discovering Another Domain of Life* (London: Oxford University Press, 2000). To my knowledge, this is the first book written exclusively about archaea, and it is written for a scientifically minded lay audience. Howland summarizes the body of scientific literature published up to 2000. It provided me with a fountain of knowledge about archaea and their peculiar habits. For a more in-depth academic perspective of archaea, I highly recommend this for further reading.

4. Günter Schäfer, Martin Engelhard, and Volker Muller, "Bioenergetics of the Archaea," *Microbiology and Molecular Biology Reviews* vol. 63, no. 3 (September 1999):570–620.

CHAPTER 2

1. The following is a collection of studies about microbial ecology and the methods that Jay Short and the scientists at Diversa, as well as scientists at the Department of Energy and academic institutions, have developed to extract raw microbial DNA and proteins from environmental samples.

Karsten Zengler, Gerardo Toledo, Michael Rappe, James Elkins, Eric J. Mathur, Jay M. Short, and Martin Keller, "Cultivating the Uncultured," *Proceedings of the National Academy of Sciences* vol. 99, no. 24 (November 26, 2002):15681–15686.

Eric Mathur, Charles Costanza, Leif Christoffersen, Carolyn Erickson, Monica Sullivan, Michelle Bene, and Jay M. Short, "An Overview of Bioprospecting and the *Diversa* Model," *IP Strategy Today* no. 11 (2004).

Susannah Green Tringe, Christian von Mering, Arthur Kobayashi, Asaf A. Salamov, Kevin Chen, Hwai W. Chang, Mircea Podar, Jay M. Short, Eric J. Mathur, John C. Detter, Peer Bork, Philip Hugenholtz, and Edward M. Rubin, "Comparative Metagenomics of Microbial Communities," *Science* vol. 308, no. 5721 (April 22, 2005):554–557.

Philip Hugenholtz, "Exploring Prokaryotic Diversity in the Genomic Era," *Genome Biology* vol. 3, no. 2 (January 29, 2002):reviews0003.1–reviews0003.8. Available online at http://genebiology.com/2002/3/2/reviews/0003.

Bess B. Ward, "How Many Species of Prokaryotes Are There?" *Proceedings of the National Academy of Sciences* vol. 99, no. 16 (August 6, 2002):10234–10236.

Allan Konopka, "Microbial Ecology: Searching for Principles, The Extraordinary Diversity of Microbial Ecosystems Complicates Efforts to Develop Principles Encompassing Microbial Ecology," *Microbe* vol. 1, no. 4 (2006).

American Academy of Microbiology, *Basic Research for the Future: Opportunities in Microbiology for the Coming Decade.* This report is based on an American Academy of Microbiology colloquium held May 4–7, 1996, in Washington, DC. The colloquium was supported by the National Science Foundation, the U.S. Department of Energy, the U.S. Department of Agriculture, and the American Society for Microbiology.

U.S. Department of Energy's Genomics: *GTL Bioenergy Research Centers White Paper.*

U.S. Department of Energy Office of Science, Office of Biological and Environmental Research Genomics: GTL Program, August 2006. Available online at http://www.genomicsgtl.energy.gov.

2. John D. Varley and Preston T. Scott, "Conservation of Microbial Diversity, A Yellowstone Priority," *ASM News* vol. 64, no. 3 (March 1998):147–151.

3. For a perspective on issues faced by indigenous people worldwide, see the documentary *Yakoana: The Voice of Indigenous Peoples*, produced by Anh Crutcher. The video can be ordered at http://www.parabola.org/.

CHAPTER 3

1. Carl Woese, "The Birth of the Archaea: A Personal Retrospective," in Roger Garrett and Hans-Peter Klenk, eds., *Archaea: Evolution, Physiology, and Molecular Biology* (Malden, MA: Blackwell Publishing, Inc., 2006).

2. Carl Woese and George Fox, "Phylogenetic Structure of Prokaryotic Domain: The Primary Kingdoms," *Proceedings of the National Academy of Sciences* vol. 74, no. 11 (November 1977):5088–5090.

3. Carl Woese, Otto Kandler, and Mark L. Wheelis, "Towards a Natural System of Organisms: Proposal for the Domains Archaea, Bacteria, and Eucarya," *Proceedings of the National Academy of Sciences* vol. 87 (June 1990):4576–4579.

4. Carl Woese, "The Archaeal Concept and the World It Lives In: A Retrospective, *Photosynthesis Research* vol. 80, nos. 1–3 (April 2004):361–372.

5. Carl Woese, "Order in the Genetic Code," *Proceedings of the National Academy of Sciences* vol. 54, no. 1 (July 1965):71–75.

6. Julie Clayton and Carina Dennis, eds., *50 Years of DNA* (Basingstoke, Hampshire, England: Nature Publishing Group, 2003). This easy-to-read book is an insightful and quite handy summary of milestones in the discovery of DNA and its contributions to modern science. I found it quite useful for describing the early history of DNA research.

7. M. Nirenberg, P. Leder, M. Bernfield, R. Brimacombe, J. Trupin, F. Rottman, and C. O'Neal, "RNA Codewords and Protein Synthesis: On the General Nature of the RNA Code," *Proceedings of the National Academy of Sciences* vol. 53 (1965):1161–1168.

8. Dieter Soll was part of the historic work on transfer RNA and codon assignments during the 1960s. In Woese's third paper published in 1965, he refers to Crick's theory if the evolution of the genetic code as the "adaptor hypothesis." But Soll says Woese is incorrect in this

attribution. As Woese states, Crick did mistakenly argue that the codon assignments are arbitrary; however, this was not Crick's adaptor hypothesis. For a better understanding of Crick's theory, including his adaptor hypothesis, see:

F.H.C. Crick, "On Protein Synthesis," *Symposia of the Society for Experimental Biology*, vol. 12 (1958):138–163.

Dieter Soll and Uttam L. RajBhandary, "The Genetic Code: Thawing the 'Frozen Accident,'" *Journal of Biosciences* vol. 31, no. 4 (October 2006):459–463.

9. King-Thom Chung and Marvin P. Bryan, "Robert E. Hungate: Pioneer of Anaerobic Microbial Ecology," *Anaerobe* vol. 3 (1997):213–217.

10. Mayr supported Woese's concept of archaebacteria until Woese decided to officially name it as a third kingdom of life. The following are references to the two papers in which the authors slugged it out in the scientific press:

Ernst Mayr, "Two Empires or Three?" *Proceedings of the National Academy of Sciences* vol. 95, no. 17 (August 18, 1998):9720–9723.

Carl Woese, "Default Taxonomy: Ernst Mayr's View of the Microbial World," *Proceedings of the National Academy of Sciences* vol. 95, no. 19 (September 15, 1998):11043–11046. Abstract excerpt: "Mayr has suggested that the now accepted classification of life into three primary domains, Archaea, Bacteria, and Eucarya—originally proposed by myself and others—be abandoned in favor of the earlier Prokaryote–Eukaryote classification. Although the matter appears a taxonomic quibble, it is not that simple. At issue here are differing views as to the nature of biological classification, which are underlain by differing views as to what biology is and will be—matters of concern to all biologists."

11. Carl Woese, "Interpreting the Universal Phylogenetic Tree," *Proceedings of the National Academy of Sciences* vol. 97, no.15 (July 18, 2000):8392–8396. Abstract excerpt: "The universal phylogenetic tree not only spans all extant life, but its root and earliest branchings represent stages in the evolutionary process before modern cell types had come into being. The evolution of the cell is an interplay between ver-

tically derived and horizontally acquired variation. Primitive cellular entities were necessarily simpler and more modular in design than are modern cells. Consequently, horizontal gene transfer early on was pervasive, dominating the evolutionary dynamic. The root of the universal phylogenetic tree represents the first stage in cellular evolution when the evolving cell became sufficiently integrated and stable to the erosive effects of horizontal gene transfer that true organismal lineages could exist."

See also:

Carl Woese, "A New Biology for a New Century," *Microbiology and Molecular Biology Reviews* vol. 68, no. 2 (June 2004):173–186.

Carl Woese, "On the Evolution of Cells," *Proceedings of the National Academy of Sciences* vol. 99, no. 13 (June 25, 2002):8742–8747.

Carl Woese, "The Universal Ancestor," *Proceedings of the National Academy of Sciences* vol. 95 (June 1998):6854–6859.

Gary J. Olsen and Carl Woese, "Ribosomal RNA: A Key to Phylogeny," *The FASEB Journal* vol. 7 (1993):113–123.

Dawn J. Brooks, Jacques R. Fresco, Arthur M. Lesk, and Mona Singh, "Evolution of Amino Acid Frequencies in Proteins Over Deep Time: Inferred Order of Introduction of Amino Acids into the Genetic Code, *Molecular Biology and Evolution* vol. 19, no. 10 (2002):1645–1655.

Berend Snel, Peer Bork, and Martijn A. Huynen, "Genomes in Flux: The Evolution of Archaeal and Proteobacterial Gene Content," *Genome Research* vol. 2, no. 1 (January 2002):17–25.

C.G. Kurland, B. Canback, and Otto G. Berg, "Horizontal Gene Transfer: A Critical View," *Proceedings of the National Academy of Sciences* vol. 100, no. 17 (August 19, 2003):9658–9662.

Nikos C. Kyrpides and Gary J. Olsen, "Archaeal and Bacterial Hyperthermophiles: Horizontal Gene Exchange or Common Ancestry?" *Trends in Genetics* vol. 15, no. 8 (August 1, 1999):298–299.

Otto G. Berg and C.G. Kurland, "Evolution of Microbial Genomes: Sequence Acquisition and Loss," *Molecular Biology and Evolution* vol. 19, no. 12 (2002):2265–2276.

Paul Schimmel and Lluis Ribas de Pouplana, "Genetic Code Origins: Experiments Confirm Phylogenetic Predictions and May Explain a Puzzle," *Proceedings of the National Academy of Sciences* vol. 96, no. 2 (January 19, 1999):327–328.

For an opposing view of Woese's determination of archaea as a third domain of life, see Radhey S. Gupta, "Protein Phylogenies and Signature Sequences: A Reappraisal of Evolutionary Relationships Among Archaebacteria, Eubacteria, and Eukaryotes," *Microbiology and Molecular Biology Reviews* vol. 62, no. 4 (December 1998):1435–1491.

CHAPTER 4

1. Having discovered and cultured the majority of species, Karl Stetter is one of the most prolific publishers of scientific papers on archaea. Rather than muddy the text with too many footnotes, I'm including here some of the papers most pertinent to the chapter.

They are listed in chronological order, but they by no means represent all of Stetter's published work from 1970 to the present:

K.O. Stetter, "Production of Exclusively L(+)-Lactic Acid Containing Food by Controlled Fermentation," *Proceedings of the First Intersectional Congress of IAMS* vol. 5 (1974):164–168.

K.O. Stetter and W. Zillig, "Transcription in Lactobacillaceae: DNA-Dependent RNA Polymerase from *Lactobacillus curvatus*," *European Journal of Biochemistry* vol. 48, no. 2 (October 1974):527–540.

R. Hensel, U. Mayr, K.O. Stetter, and O. Kandler, "Comparative Studies of Lactic Acid Dehydrogenases in Lactic Acid Bacteria," *Archives of Microbiology* vol. 112, no. 1 (February 1977):81–93.

K.O. Stetter, W. Zillig, and M. Tobien, "DNA-Dependent RNA Polymerase from Halobacterium Halobium," *European Journal of Biochemistry* vol. 91 (1978):193–199.

W. Zillig, K.O. Stetter, and D. Janekovic, "DNA-Dependent RNA Polymerase from the Archaebacterium Sulfolobus Acidocaldarius," *European Journal of Biochemistry* vol. 96 (1979):597–604.

W. Zillig, K.O. Stetter, W. Schulz, and D. Janekovic, "Comparative Studies of Structure and Function of DNA-Dependent RNA Polymerases from Eubacteria and Archaebacteria," in P. Mildner and B. Ries, eds., *Trends in Enzymology: Proceedings of the*

Federation of European Biochemical Societies Meeting on Enzymes vol. 60 (UK: Pergamon Press, 1980), 159–178.

W. Zillig, K.O. Stetter, S. Wunderl, W. Schulz, H. Priess, and I. Scholz, "The Sulfolobus-"Caldariella" Group: Taxonomy on the Basis of the Structure of DNA-Dependent RNA Polymerases," *Archives of Microbiology* vol. 125, no. 3 (April 1980):259–269.

S. Sturm, U. Schönefeld, W. Zillig, D. Janekovic, and K.O. Stetter, "Structure and Function of the DNA-Dependent RNA Polymerase of the Archaebacterium Thermoplasma Acidophilum," *Zentralblatt für Bakteriologie, Mikrobiologie und Hygiene, Abt. 1, Originale C: Allgemeine, Angewandte und Oekologische Mikrobiologie* vol. 1 (1980):12–25.

W. Zillig and K.O. Stetter, "Distinction Between the Transcription Systems of Archaebacteria and Eubacteria," in S. Osawa, H. Ozeki, H. Uchida, and T. Yura, eds., *Genetics and Evolution of RNA Polymerase, tRNA and Ribosomes* (University of Tokyo Press and Elsevier, 1980):525–538.

K.O. Stetter, J. Winter, and R. Hartlieb, "DNA-Dependent RNA Polymerase of the Archaebacterium Methanobacterium Thermoautotrophicum," *Zentralblatt für Bakteriologie, Mikrobiologie und Hygiene, Abt. 1, Originale C: Allgemeine, Angewandte und Oekologische Mikrobiologie* vol. 1 (1980):201–214.

K.O. Stetter, M. Thomm, J. Winter, G. Wildgruber, H. Huber, W. Zillig, D. Janecovic, H. König, P. Palm, and S. Wunderl, "Methanothermus fervidus, sp. nov., a Novel Extremely Thermophilic Methanogen Isolated from an Icelandic Hot Spring," *Zentralblatt für Bakteriologie, Mikrobiologie und Hygiene, Abt. 1, Originale C: Allgemeine, Angewandte und Oekologische Mikrobiologie* vol. 2 (1981):166–178.

W. Zillig, K.O. Stetter, W. Schäfer, D. Janekovic, S. Wunderl, I. Holz, and P. Palm, "Thermoproteales, a Novel Type of Extremely Thermoacidophilic Anaerobic Archaebacteria Isolated from Icelandic Solfataras," *Zentralblatt für Bakteriologie, Mikrobiologie und Hygiene, Abt. 1, Originale C: Allgemeine, Angewandte und Oekologische Mikrobiologie* vol. 2 (1981):205–227.

H. Huber, M. Thomm, H. König, G. Thies, and K.O. Stetter, "Methanococcus thermolithotrophicus, a Novel Thermophilic Lithotrophic Methanogen," *Archives of Microbiology* 132, no. 1 (July 1982):47–50.

W. Zillig, K.O. Stetter, R. Schnabel, J. Madon, and A. Gierl, "Transcription in Archaebacteria," *Zentralblatt für Bakteriologie, Mikrobiologie und Hygiene, Abt. 1, Originale C: Allgemeine, Angewandte und Oekologische Mikrobiologie* vol. 3 (1982):218–227.

K.O. Stetter, H. König, and E. Stackebrandt, "Pyrodictium gen. nov., a New Genus of Submarine Disc-Shaped Sulphur Reducing Archaebacteria Growing Optimally at 105°C," *Systematic and Applied Microbiology* vol. 4, no. 4 (1983):535–551.

K.O. Stetter, "Anaerobic Life at Extremely High Temperatures," *Origins of Life and Evolution of the Biosphere* vol. 14, nos. 1–4 (December 1984):809–815.

W. Zillig, R. Schnabel, F. Gropp, W.D. Reiter, K. Stetter, and M. Thomm, "The Evolution of the Transcription Apparatus," in K.H. Schleifer and E. Stackbrandt, eds., *Evolution of Prokaryotes* (New York: Academic Press, 1985), 45–72.

K.O. Stetter, "Thermophilic Archaebacteria Occurring in Submarine Hydrothermal Areas," in D. Caldwell and C. Brierley, eds., *Planetary Ecology* (New York: Van Nostrand Reinhold, 1985), 320–322.

R. Huber, T.A. Langworthy, H. König, M. Thomm, C.R. Woese, U.B. Sleytr, and K.O. Stetter, "*Thermotoga maritima* sp. nov. Represents a New Genus of Unique Extremely Thermophilic Eubacteria Growing up to 90°C," *Archives of Microbiology* vol 144, no. 4 (May 1986):324–333.

G. Fiala and K.O. Stetter, "*Pyrococcus furiosus* sp. nov. Represents a Novel Genus of Marine Heterotrophic Archaebacteria Growing Optimally at 100°C," *Archives of Microbiology* vol. 145, no. 1 (June 1986):56–61.

K.O. Stetter, A. Segerer, W. Zillig, G. Huber, G. Fiala, R. Huber, and H. König, "Extremely Thermophilic Sulfur-Metabolizing Archaebacteria," *Systematic and Applied Microbiology* vol. 7, nos. 2–3 (1986):393–397.

K.O. Stetter, G. Fiala, R. Huber, G. Huber, and A. Segerer, "Life Above the Boiling Point of Water?" *Cellular and Molecular Life Sciences* vol. 42, nos. 11–12 (December 1986):1187–1191.

R. Huber, M. Kurr, H.W. Jannasch, and K.O. Stetter, "A Novel Group of Abyssal Methanogenic Archaebacteria (Methanopyrus) Growing at 110°C," *Nature* vol. 342 (December 14, 1989):833–834.

R. Huber, T. Wilharm, D. Huber, A. Trincone, S. Burggraf, H. König, R. Rachel, I. Rockinger, H. Fricke, and K.O. Stetter, "Aquifex pyrophilus gen. nov. sp. nov. Represents a Novel Group of Marine Hyperthermophilic Hydrogen-Oxidizing Bacteria," *Systematic and Applied Microbiology* vol. 15, no. 3 (1992):340–351.

K.O. Stetter, R. Huber, E. Blöchl, M. Kurr, R.D. Eden, M. Fielder, H. Cash, and I. Vance, "Hyperthermophilic Archaea Are Thriving in Deep North Sea and Alaskan Oil Reservoirs," *Nature* vol. 365 (October 21, 1993):743–745.

A.H. Segerer, S. Burggraf, G. Fiala, G. Huber, R. Huber, U. Pley, and K.O. Stetter, "Life in Hot Springs and Hydrothermal Vents," *Origins of Life and Evolution of the Biosphere* vol. 23, no. 1 (February 1993):77–90.

K.O. Stetter, "The Lesson of Archaebacteria," in S. Bengston, ed., *Early Life on Earth*, Symposium No. 84 (New York: Columbia University Press, 1994), 143–151.

K.O. Stetter, "Hyperthermophiles in the History of Life," in G.R. Block and J.A. Goode, eds., *Evolution of Hydrothermal Ecosystems on Earth (and Mars?)*, CIBA Foundation Symposium No. 202 (New York: John Wiley and Sons, 1996), 1–18.

E. Blöchl, R. Rachel, S. Burggraf, D. Hafenbradl, H.W. Jannasch, and K.O. Stetter, "Pyrolobus fumarii, gen. and sp. nov. Represents a Novel Group of Archaea, Extending the Upper Temperature Limit for Life to 113°C," *Extremophiles* vol. 1, no. 1 (February 1997):14–21.

H. Huber and K.O. Stetter, "Hyperthermophiles and Their Possible Potential in Biotechnology," *Journal of Biotechnology* vol. 64 (1998):39–52.

K.O. Stetter, "Hyperthermophiles and Their Possible Role as Ancestors of Modern Life," in A. Brack, ed., *The Molecular Origins of Life: Assembling Pieces of the Puzzle* (New York: Cambridge University Press, 1999), 315–335.

Harald Huber, Siegfried Burggraf, Thomas Mayer, Irith Wyschkony, Reinhard Rachel, and Karl O. Stetter, "Ignicoccus gen. nov., a Novel Genus of Hyperthermophilic, Chemolithoautotrophic Archaea, Represented by Two New Species, Ignicoccus islandicus sp. nov. and Ignicoccus pacificus sp. nov," *International Journal of Systematic and Evolutionary Microbiology* vol. 50 (2000):2093–2100.

K.O. Stetter and R. Huber, "The Role of Hyperthermophilic Prokaryotes in Oil Fields," in C.R. Bell, M. Brylinsky, and P. Johnson-Green, eds., *Microbial Biosystems: New Frontiers*, Proceedings of the 8th International Symposium on Microbial Ecology (Halifax: Atlantic Canada Society for Microbial Ecology, 2000), 369–375.

K.O. Stetter, "Hyperthermophilic Microorganisms," in G. Horneck and C. Baumstark-Khan, *Astrobiology: The Quest for the Conditions of Life* (New York: Springer Verlag, 2002), 169–184.

H. Huber, M.J. Hohn, R. Rachel, T. Fuchs, V.C. Wimmer, and K.O. Stetter, "A New Phylum of Archaea Represented by a Nanosized Hyperthermophilic Symbiont," *Nature* vol. 417 (May 2, 2002):63–67.

K.O. Stetter, M.J. Hohn, B.P. Hedlund, and H. Huber, "Detection of 16S rDNA Sequences Representing the Novel Phylum 'Nanoarchaeota': Indication for a Wide Distribution in High Temperature Biotopes," *Systematic and Applied Microbiology* vol. 25, no. 4 (December 2002):551–554.

E. Waters, M.J. Hohn, I. Ahel, D.E. Graham, M.D. Adams, M. Barnstead, K.Y. Beeson, L. Bibbs, R. Bolanos, M. Keller, K. Kretz, X. Lin, E. Mathur, J. Ni, M. Podar, T. Richardson, G.G. Sutton, M. Simon, D. Söll, K.O. Stetter, J.M. Short, and M. Noordewier, "The Genome of Nanoarchaeum Equitans: Insights into Early Archaeal Evolution and Derived Parasitism," *Proceedings of the National Academy of Sciences* vol. 100, no. 22 (October 28, 2003):12984–12988.

K.O. Stetter, "Hyperthermophiles: Microbes at the Upper Temperature Border of Life," Professor Will N. Koning's Laudatio and Professor Stetter's Antonie van Leeuwenhoek Award Lecture, *ASM News* vol. 70, no. 5 (May 2004):247–248.

K.O. Stetter, "Volcanoes, Hydrothermal Venting, and the Origin of Life," in J. Marti and G. G. J. Ernst, eds., *Volcanoes and the Environment* (New York: Cambridge University Press, 2005), 175–206.

K.O. Stetter, M.J. Hohn, H. Huber, R. Rachel, E. Mathur, B. Hedlund, and U. Jahn, "A Novel Kingdom of Parasitic Archaea," in W.P. Inskeep and T.R. McDermott, eds., *Geothermal Biology and Geochemistry in Yellowstone National Park*, Proceedings of the ThermalBiology Institute Workshop (Bozeman: Montana State University Publications, 2005), 249–259.

2. Thomas Brock, "Microbiology in Yellowstone at First Focused on the Basic Science and Ecology but Gradually Has Expanded in Scope," *ASM News* vol. 64, no. 3 (1998):137–140.

3. Ralph S. Wolfe, "Anaerobic Life: A Centennial View," *Journal of Bacteriology* vol. 181, no. 11 (June 1999):3317–3320.

4. L.A. Morgan, W.C. Shanks, III, D.A. Lovalvo, S.J. Johnson, W.J. Stephenson, K.L. Pierce, S.S. Harlan, C.A. Finn, G. Lee, M. Webring, B. Schulze, J. Duhn, R. Sweeney, and L. Balistrieri, "Exploration and Discovery in Yellowstone Lake: Results from High-Resolution Sonar Imaging, Seismic Reflection Profiling and Submersible Studies," *Journal of Volcanology and Geothermal Research* vol. 122, no. 3 (April 1, 2003):221–242.

5. L.A. Morgan, W.C. Shanks, III, D. Lovalvo, G. Lee, M. Webring, W.J. Stephenson, and S.Y. Johnson, "The Floor of Yellowstone Lake Is Anything but Quiet: New Discoveries from High-Resolution Sonar Imaging, Seismic Reflection Profiling and Submersible Studies," *Yellowstone Science* vol. 11, no. 2 (Spring 2003):14–30.

6. Arthur Ashkin, "Optical Trapping and Manipulation of Neutral Particles Using Lasers," *Proceedings of the National Academy of Sciences* vol. 94 (May 1997):4853–4860.

CHAPTER 5

1. William H. Garzke Jr., David K. Brown, Paul K. Matthias, Roy Cullimore, David Wood, David Livingstone, H.P. Leighly Jr., Timothy Foecke, and Arthur Sandiford, "*Titanic*: The Anatomy of a Disaster," A Report from the Marine Forensic Panel (SD-7,. published in the proceedings of the 1997 Annual Meeting of the Society of Naval Architects and Marine Engineers held in Ottawa, Canada.

2. Charles Pellegrino and Roy Cullimore, "The Rebirth of the RMS *Titanic*: A Study of the Bioarcheology of a Physically Disrupted Sunken Vessel," *Voyage* no. 25 (June 1997):39–46.

3. Much of the material for this chapter is culled from a long telephone interview with Roy Cullimore, as well as from the book Ghosts of the Titanic by Charles Pellegrino and Roy Cullimore. But Cullimore also published several titles about the rusticles with his colleague Lori Johnston. The following is a list of publications relevant to rusticles:

G. Alford and Roy Cullimore, *The Application of Heat and Chemicals in the Control of Biofouling Events in Wells*, Sustainable Wells series (CRC Press, 1998), 162–166.

R. Cullimore and L. Johnston, "The Fate of the Iron: More Lessons from the Titanic Tragedy," *Maritime Reporter* (August 1999):22, 66.

R. Cullimore and L. Johnston, "The Science and the RMS *Titanic*: The Biological Odyssey," *Voyage* no. 32 (2000):172–176.

R. Cullimore and L. Johnston, "The Impact of Bioconcretious Structures (Rusticles) on the RMS *Titanic*: Implications to Maritime Steel Structures," The Society of Naval Architects and Marine Engineers Annual Conference, Vancouver, British Columbia, Canada, presentation October 6, 2000.

R. Cullimore and L. Johnston, "Biodeterioration of the RMS *Titanic*," *L'Actualité Chimique Canadienne* (November/December 2000):14–15.

For more on Cullimore, see his Web site at http://www.dbi.sk.ea.

4. Mitchell L. Sogin, Hilary G. Morrison, Julie A. Huber, David Mark Welch, Susan M. Huse, Phillip R. Neal, Jesus M. Arrieta, and Gerhard J. Herndl, "Microbial Diversity in the Deep Sea and the Underexplored 'Rare Biosphere,'" *Proceedings of the National Academy of Sciences* vol. 103, no. 32 (August 8, 2006):12115–12120.

For more on rusticles, review Carolyn Fry, "Iron Rations," *NewScientist* no. 2405 (July 26, 2003).

CHAPTER 6

1. Yellowstone Park Foundation, "A Proof-of-Concept Molecular All-Taxa Biodiversity Inventory (MATBI) in Yellowstone National Park: Microbial and Small Metazoan Diversity Survey in Yellowstone Lake

Using Ribosomal Gene Sequences," A report submitted by John Varley and Diversa Corp. to The Gordon and Betty Moore Foundation as stipulated by a grant made in June 2005. For a summary of the MATBI results see:

John D. Varley, "Molecular Approach to Biodiversity Inventory Doubles List of Known *Species in Yellowstone Lake, National Park System Natural Resource Year in Review*—2005. Available online at http://www2.nature.nps.gov/yearinreview/PDF/YIR2005_03.pdf.

David M. Ward, "Microbiology in Yellowstone National Park," *ASM News* vol. 64, no. 3 (March 1998):141–146.

Alice Wondrak Biel with Lisa Morgan, "New Discoveries on Yellowstone Lake's Floor," *National Park System Natural Resource Year in Review—2003.*

2. One of the greatest sins a science writer can make is the sin of omission. I have discussed Norm Pace's work throughout this book, but I have not gone into as much detail as I would have liked. Pace also is a prolific publisher of scientific articles on the topic of archaea, bacteria, and microbial ecology. The following is a partial list of some of Pace's review articles, which encapsulate much of his work:

D. Evans, S.M. Marquez, N.R. Pace, "RNase P: Interface of the RNA and Protein Worlds," *Trends in Biochemical Sciences* vol. 31, no. 6 (June 2006):333–341.

C.E. Robertson, J.K. Harris, J.R. Spear, N.R. Pace, "Phylogenetic Diversity and Ecology of Environmental Archaea," *Current Opinion in Microbiology* vol. 8, no. 6 (December 2005):638–642.

E.F. DeLong and N.R. Pace, "Environmental Diversity of Bacteria and Archaea," *Systemic Biology* vol. 50, no. 4 (August 1, 2001):470–478.

N.R. Pace, "A Molecular View of Microbial Diversity and the Biosphere," *Science* vol. 276, no. 5313 (May 2, 1997):734–740.

P. Hugenholtz and N.R. Pace, "Identifying Microbial Diversity in the Natural Environment: A Molecular Phylogenetic Approach," *Trends in Biotechnology* vol. 14, no. 6 (June 1996):190–197.

N.R. Pace, "Opening the Door onto the Natural Microbial World: Molecular Microbial Ecology," *Harvey Lectures 1995–1996* vol. 91:59–78.

N.R. Pace, "Origin of Life: Facing up to the Physical Setting," *Cell* vol. 65, no. 4 (May 17, 1991):531–533.

K.G. Field, G.J. Olsen, D.J. Lane, S.J. Giovannoni, M.T. Ghiselin, E.C. Raff, N.R. Pace, and R.A. Raff, "Molecular Phylogeny of the Animal Kingdom," *Science* vol. 239, no. 4841 (February 12, 1988):748–753.

N.R. Pace, "Structure and Synthesis of the Ribosomal Ribonucleic Acid of Prokaryotes," *Bacteriological Reviews* vol. 37, no. 4 (December 1973):562–603.

3. Roger J. Anderson and David Harmon, "Yellowstone Lake: Hotbed of Chaos or Reservoir of Resilience?" Proceedings of the Sixth Biennial Conference on the Greater Yellowstone Ecosystem (2002).

See also:

James S. Maki, Carl M. Schroeder, James C. Bruckner, Charles Wimpee, Andrew Weir, Charles C. Remsen, Carmen Aguilar, and Russell L. Cuhel, "Investigating the Microbial Ecology of Yellowstone Lake," Proceedings of the Sixth Biennial Conference on the Greater Yellowstone Ecosystem (2002).

Russell L. Cuhel, Carmen Aguilar, Charles C. Remsen, James S. Maki, David Lovalvo, J. Val Klump, and Robert W. Paddock, "The Bridge Bay Spires: Collection and Preparation of a Scientific Specimen and Museum Piece," Proceedings of the Sixth Biennial Conference on the Greater Yellowstone Ecosystem (2002).

CHAPTER 7

1. In addition to the August 2006 issue of *Elements* mentioned in the main text, the following is a list of papers that discuss zircon crystal dating. These papers somehow missed the attention of the general media, but they provided the foundation for a whole new understanding of early Earth.

Aaron J. Cavosie, Simon A. Wilde, Dunyi Liu, Paul W. Weiblen, and John W. Valley, "Internal Zoning and U–Th–Pb Chemistry of Jack Hills Detrital Zircons: A Mineral Record of Early Archean to Mesoproterozoic (4348–1576 Ma) Magmatism," *Precambrian Research* vol. 135 (2004):251–279.

A.J. Cavosie, J.W. Valley, and S.A. Wilde, "Magmatic _^{18}O in 4400–3900 Ma Detrital Zircons: A Record of the Alteration and Recycling of Crust in the Early Archean," *Earth and Planetary Science Letters* vol. 235 (2005):663–681.

H.J. Van Es, D.I. Vainshtein, R.J. De Meijer, H.W. Den Hartog, J.F. Donoghue, and A. Rozendaal, "Mineral Zircon: A Novel Thermoluminescence Geochronomoter," *Radiation Effects and Defects in Solids* vol. 157, no. 6 (2002):1063–1070.

Another must read is this paper on evidence of early life: Peter W.U. Appel, Christopher M. Fedo, Stephen Moorbath, and John S. Myers, "Early Archaean Isua Supracrustal Belt, West Greenland: Pilot Study of the Isua Multidisciplinary Research Project," *Geology of Greenland Survey Bulletin* vol. 180 (1998):94–99.

2. Ricardo Cavicchioli, "Cold-Adapted Archaea," *Nature Reviews Microbiology* vol. 4 (May 2006):331–343.

CHAPTER 8

1. More information on JGI's sequencing projects involving microbes can be found online at http://www.jgi.doe.gov/sequencing. I spent many hours surfing JGI Web pages. The Web site http://www.jgi.doe.gov has information on JGI and microbial genomics.

Much of what I learned for this chapter came from numerous face-to-face interviews with most of the principles, in particular Jay Short, Craig Venter, and Eddie Rubin. I cited directly in the text most of the relevant research, but I've listed here a few more papers worth checking out.

Jay Short has authored hundreds of scientific articles. Several key papers are cited throughout the text. But here are his two historic papers on cloning vectors:

Steven W. Kohler, G. Scott Provost, Patricia L. Kretz, Mark J. Dycaico, Joseph A. Sorge and Jay M. Short, "Development of a Short-Term, in Vivo Mutagenesis Assay: The Effects of Methylation on the Recovery of a Lambda Phage Shuttle Vector from Transgenic Mice," *Nucleic Acids Research* vol. 18, no. 10 (May 25, 1990):3007–3013.

Jay M. Short, Joseph M. Fernandez, Joseph A. Sorge, and William D. Huse, "λZAP: A Bacteriophage X Expression Vector with in Vivo Excision Properties," *Nucleic Acids Research* vol. 16, no. 15 (1998):7583–7600.

Craig Venter's list of publications is too numerous to cite, but it turns out that Venter is the topic of an entire course in biotechnology. Here is the Web page for the course syllabus with links to many of Venter's key publications: http://www.biology.iupui.edu/biocourses/Biol540/15genomefull2k6.html. To find more about Venter's current work, see http://www.venterinstitute.org.

2. What follows is a list of materials authored or coauthored by scientist Ken Nealson, Wrigley Professor of Geobiology at the University of Southern California. I discovered Nealson when researching shewanella, unfortunately at a time when it was too late to go back and include more of his research in this book, which would had necessitated adding another 30 pages to the text.

National Research Council, *Biological Contamination of Mars: Issues and Recommendations* (Washington, DC: National Academy Press, 1992).

National Research Council, *Mars Sample Return: Issues and Recommendations* (Washington, DC: National Academy Press, 1997).

"The Limits of Life on Earth and Searching for Life on Mars," *Journal of Geophysical Research* vol. 102, no. 23 (October 25, 1997):675–686.

"Nannobacteria: Size Limits and Evidence," *Science* vol. 276, no. 5320 (June 20, 1997):1773–1776.

"Post-Viking Microbiology: New Approaches, New Data, New Insights," *Origins of Life and Evolution of Biospheres* vol. 29, no. 1 (January 1999):73–93.

"The Search for Extraterrestrial Life," *Engineering & Science* vol. 62, no.1/2 (1999):30–39.

E.J. Gaidos, K.H. Nealson, and J.L. Kirschvink, "Life in Ice-Covered Oceans," *Science* vol. 284, no. 5420 (June 4, 1999):1631–1633.

K.H. Nealson and P.G. Conrad, "Life: Past, Present and Future," *Philosophical Transactions of the Royal Society B: Biological Science* vol. 354, no. 1392 (December 29, 1999):1923–1939.

B.P. Weiss, Y. Yung, and K.H. Nealson, "Atmospheric Energy for Subsurface Life on Mars?" *Proceedings of the National Academy of Sciences* vol. 97, no. 4 (February 15, 2000):1395–1399.

P.G. Conrad and K.H. Nealson, "A Non-Earthcentric Approach to Life Detection," *Astrobiology* vol. 1, no. 1 (March 2001):15–24.

See also Nealson's Web site at http://genomicsgtl.energy.gov/research/shewanella/ken.pdf. O particular interest is his PowerPoint presentation titled "SHEWANELLA and Genomes to Life!! THE FUTURE!! WHERE ARE WE GOING? HOW WILL WE GET THERE? WHAT ARE THE CHALLENGES AND TRAPS?"

ACKNOWLEDGMENTS

So many people provided the knowledge and insights to undertake a project that is, to put it mildly, far over my head as a non-scientist. Many others made life bearable and even joyous during the struggles to complete this project. Jeffrey Robbins, I want to thank you first for making this book possible and for having such a kind heart. I couldn't have asked for a better editor and mentor on this project.

I also owe much to my daughters Amber and Kelsey for always wanting more money (and of course for returning so much joy), and to Mom, who called every day to ask how the manuscript was coming along. You always provide comfort and do your best to keep me on track. Leigh, tiny dancer, you are the inspiration in everything. Vous êtes la lumière du soleil dans mon univers.

Carl Woese, I believe you are the most remarkable microbiologist of our time. Secrets to the origin of life are at your fingertips. The world will remember your achievements. Thank you for your patience and trust. Karl Stetter, you taught modern microbiology "culture." Thank you for your generosity and wisdom throughout this project.

Without your teaching, this reporter wouldn't have had a chance at even trying to explain something as complicated as the archaea and microbial metabolism.

Eric Mathur, scorpion brother, diving buddy, and partner in crime, you first told me the story of archaea and then encouraged me to jump. Not sure whether to shoot you or thank you, but none of this would have happened without your inspiration. Jay Short, you taught me to free-fall and then to fly. As you said, "Success is falling down seven times, but getting up eight." You made the transition possible and you've become a true friend. Your creativity and scientific achievements have transformed modern microbial ecology and biotechnology. Leif Christoffersen, you are the soldier—loyal, brave, and ready to take the bullets when they fly. Next time we go to into the heart of darkness, you're coming with me. Steve Briggs taught me about soil bacteria and many intricacies of the microbial universe, and then I'm sure I felt your hand on my back when Eric shouted "Jump!" The rest of the Diversa Corporation—Ed Shonsey, all the scientists and staff, including Paula and Lisa—thanks for trusting me with your secrets, allowing me into your world, and giving me a home.

Roy Cullimore, I believe you are one of the most insightful scientists on the planet. Your work with the rusticles taught me that microbes are much more than the sum of their parts. You pointed me to the clouds, literally. Charles Pellegrino, your writing and vision of the rusticles were a tremendous inspiration. Thank you for introducing me to the entity you aptly named *thing*.

Dennis Kelly, Linda Kauss, and Susan Weiss of *USA Today* allowed me to explore the world for 16 years and provided the most remarkable patience, kindness, and support I have known. (Dennis, you are my real "Dad.") Dan Vergano carried the load when I was dragging, then "stole" my job and kept me plugged into the old world when I was free-falling. (We always knew Dad liked you better!) Bob Davis, what can I say? Your support as a friend and colleague are immeasurable. (Running in the sleet along the Potomac: priceless.)

I want to thank Craig Venter for 15 years of friendship and for showing me the future through your incredible crystal ball. You sparked my lust for adventure aboard *Sorcerer* and taught me to overcome fear. You showed me that it's not about plan A. It's how you execute plans B, C, and D. (Maybe someday they'll let us return to Foxy's.)

Eddy Rubin, thank you for your guidance, wisdom, the best coffee in the world, and inspiring me to want to surf. Ed DeLong, you were looking in all of the right places for the archaea. You showed everyone that they are everywhere.

Norm Pace and Chuck Robertson—you know what you did. Thank you for saving me from myself, and Norm especially, thank you for your direction. You are a great teacher with a big heart.

Jeff Robbins, you have the patience of Job. Thank you for coaching me to the finish line. Jennifer Gates, as always, you are the matchmaker from Heaven. Julia Abrahams, your suggestions were solid and let's hope they saved me from too much embarrassment.

Jim Barlow and Roger Segelken, two of the best science writers in the business, introduced me to Carl Woese and Tom Gold. You guys did all the hard work over the years that reporters steal and like to call their own. You knew the importance of archaea long before the rest of us.

John Varley, Yellowstone's rock of a scientist and a truly kind gentleman, you are the model that anyone who cares about conservation should follow. We will figure out the nature of those spires before it's over.

John Valley is, in my opinion, the number one geologist who discovered life on Earth. How in the heck did you and your colleagues figure out how to unlock 5 billion years of history in a tiny zircon crystal? Your August 2006 issue of *Elements* and the authors are reshaping everything we thought we knew about our planet in its infancy. Your group's careful work has been one of the greatest inspirations in my 20 years as a science writer.

I wish to acknowledge all of the scientists whose work is cited throughout the manuscript. I apologize to all whose names and institutions have been omitted. They were there along with your citations in the original version! Honest. And thank you to all of the public information officers at universities in the United States, Europe, and Australia, whose work I have relied upon, and a special thanks to NASA, the Department of Energy, and the National Science Foundation.

Finally, to Father Kleinmann and Father Hanley: It's a mystery!

INDEX

A

B

C

F

G

N

O

P

Q

R

S

U

V

Z